D0849334

MORAL ORIGINS

ALSO BY CHRISTOPHER BOEHM

Hierarchy in the Forest

Blood Revenge

Montenegrin Social Organization and Values

MORAL ORIGINS

The Evolution of Virtue, Altruism, and Shame

CHRISTOPHER BOEHM

BASIC BOOKS
A MEMBER OF THE PERSEUS BOOKS GROUP
NEW YORK

Published by Basic Books,
A Member of the Perseus Books Group

Books published by Basic Books are available at special
discounts for bulk purchases in the United States by corporations,
institutions, and other organizations. For more information, please contact
the Special Markets Department at the Perseus Books Group, 2300 Chestnut
Street, Suite 200, Philadelphia, PA 19103, or call (800) 810-4145,
ext. 5000, or e-mail special.markets@perseusbooks.com.

Designed by Linda Mark
Text set in 11 point Giovanni by the Perseus Books Group

Library of Congress Cataloging-in-Publication Data
Boehm, Christopher.
Moral origins : the evolution of virtue, altruism,
and shame / Christopher Boehm.
p. cm.
Includes bibliographical references and index.
ISBN 978-0-465-02048-5 (hardcover : alk. paper)
ISBN 978-0-465-02919-8 (e-book)
1. Ethics, Evolutionary. 2. Virtue. 3. Altruism. 4. Shame. I. Title.
BJ1311.B645 2012
155.7--dc23
2011048896

2 1

This book is dedicated to
the memory of Donald T. Campbell

CONTENTS

1 DARWIN'S INNER VOICE

A NATURAL-BORN HERESY

Queen Victoria's England provided a most comfortable environment for Christians who loved to take their Bibles literally. Nature was perfect because in just seven days God had made nature perfect. Oceans and fish, predators and prey, all fit together like hands in gloves. And this perfectly tuned natural world was forever fixed and static because Jehovah's unlimited powers had made it that way.[1]

Not only that, but the Old Testament's Adam and Eve were real, if uniquely special people whose divine Maker had created them quite recently. A methodical churchman had actually done the biblical math and concluded that just under six thousand years had elapsed since God had made Eve from Adam's rib, installed this first pair of humans in His idyllic Garden of Eden, and then left them to their fate. In terms of evolutionary time, this meant that the origin of fallible human choice and our sinful sense of shame—together, these gave us a conscience—had taken place only yesterday or perhaps the day before. However, in the twenty-second year of the devout queen's reign all of this was about to change—and for many there would be no turning back.

In 1859, Charles Darwin's *On the Origin of Species* shook cultured reading publics in Great Britain and elsewhere like an irreverent clap of thunder.[2] The initial Darwinian lightning bolt did not strike directly against the sacred moral origins story of Eve, Adam, and the persuasive serpent who seemed to undermine the good works of an otherwise omnipotent Jehovah. Rather, this new scientific thesis introduced to the physical world of animals and plants a theory of gradual but ever-changing transformations that were wholly naturalistic. As a result, the beautiful fit of species to their environments was no longer the work of God; indeed, the banal process of natural selection operated very much as livestock breeders did when with short-term practical objectives they changed the hereditary destinies of the animals they domesticated.

These breeders did their work in a deliberate fashion. They permitted more favorably endowed individuals to flourish, while denying less useful or less aesthetically pleasing individuals the opportunity to reproduce. Robert Darwin, a doctor and a country gentleman, was one of these animal breeders. His thoughtful son—who as a young man seemed to be headed for the ministry—knew that individuals of a domesticated species varied along many dimensions. Cattle varied in their productivity in giving milk; dogs, in their natural tendencies to point and fetch, their degree of docility, and the color of their coats.[3] It was only after years of sailing around the world as a professional naturalist that the young Charles Darwin became aware that nondomesticated species were just as variable as their domesticated brethren.[4]

Darwin's official job had been to collect museum samples and to describe in minute detail the species of plants and animals on different continents, and all this hard work led, as we know, to a major theory. In his mind, such hereditary variation was something that a *natural* type of selection could act upon spontaneously: the individuals who were more fit to deal with their environments could reproduce, multiply, and flourish, while those who weren't, couldn't. This singular insight was to change the Western world's notions about nature and even its larger view of the universe.

This brings us to a profound difference between natural selection and the selection practiced by animal breeders. For Darwin, changeable natural environments were doing the mechanical work of triage,[5] and unlike either a practical livestock breeder or a purposeful God Almighty, these environments had no intentions whatsoever. They were acting as *blind* arbiters,[6] rather than as deliberate agents who knew what they were doing, and this meant that the perfection of nature was just one big accident. The specter loomed of a scary world, devoid of any ultimate Purpose, that suddenly lacked a protective, omniscient, and omnipotent God whose comforting role it was to help those who faithfully prayed for His assistance.

After the passage of a century and a half, it's remarkable for any major theory not to be superseded, or at least vastly modified. However, in its basics this blind, mechanical theory of natural selection is still going strong in the world of science.[7] If we add "genes" to what Darwin thought of rather intuitively as hereditary variation, the idea of natural environments favoring some variants and selecting against others works just as well in the early twenty-first century as it did in the mid-nineteenth. When we consider the complexities of life processes, the simplicity and explanatory power of the theory are awesome.

INDIVIDUAL COMPETITION DID THE TRICK

As the title tells us, *On the Origin of Species* was about how species come into being naturally—that is, without any supernatural help. To illustrate Darwin's thinking, let's consider a hypothetical. If a primitive bear species had been distributed originally over a restricted, uniform part of North America and then portions of this population began to migrate into adjacent areas, as their gene pools became separated, these bear subpopulations might gradually begin to differ because they were coping with different climates or new food sources—until eventually some of them could no longer interbreed. The result might be something like what we have today: black bears, brown bears (including grizzlies), and polar bears.

Darwin's real-life examples of speciation were taken from all manner of animals and plants, large and small, and in the absence of DNA analysis his scientific stories about the selection pressures that shaped these organisms rang of impeccable logic and sound scholarship. In modern terms, what Darwin's theory told the world was that potentially changeable natural environments were acting continuously on variation in the gene pools of resident populations. And for this process to do its work, two conditions were needed mechanically: hereditary variation and finite life spans among individuals. The latter was necessary if gene pools were to be modified over generations, for the less fit had to fade away and be replaced if local populations were to evolve to resemble their fittest members.

A special insight about the breeding potential of animals guided Darwin's theorizing. He founded his entire theory of natural selection on one simple but staggering mathematical realization of late-eighteenth-century English political economist and demographer the Reverend Thomas Robert Malthus:[8] if all living organisms were to procreate to their full capacity (think of dogs and cats with their annual litters), in theory within not too many generations the planet would be so exponentially overpopulated that there would be little to eat and, eventually, literally nowhere to stand. Darwin's beautifully simplified answer to a planet knee-deep in living things was sloganized by evolutionary sociologist Herbert Spencer as the "survival of the fittest,"[9] even though Darwin's theory was more subtle than that.

If the systems of biological evolution that Darwin described so mechanically were devoid of guidance from on high, then somehow they had to "regulate" themselves. As an illustration, if a population became denser, food would become scarcer and indirect competition based on who was more efficient at foraging or predation would intensify. At some point this would limit the population's growth, and its size would stabilize and remain in equilibrium. It was thus that

Darwin's theory solved Malthus's problem of potentially unlimited and exponential population growth.

Darwin's new ideas challenged not only God's directly manipulative role as portrayed in the Book of Genesis, but also the timeline of Creation. Darwin thought of biological evolutionary processes as being very gradual, and in doing so, he was able to take cues from a different field of study, namely, geology. Increasingly, naturalists like Scottish lawyer and geologist Charles Lyell had been hypothesizing that geological formations changed very gradually over time,[10] owing to the action of water or wind. Religious skeptics were coming to understand that these processes had required not a few thousand but millions of years to take place. As geological cues helped Darwin to realize that the various physical landscapes he scrutinized in his travels were far from static, this provided yet another essential ingredient for his dynamic but gradualistic theory of environmentally driven natural selection and species origination.

Thus, the biblical story of instant and permanent Creation was being undermined on a number of fronts, all unified by Darwin into an extremely logical—and exquisitely documented—theory of natural selection. His new theory so rudely challenged the static beliefs of religious fundamentalists that many of them were aroused to personally denounce "evolutionism," like today's antiscientific religious believers who try to pick apart the scenarios that evolutionists create and publish. They often assume that a few heretofore-unexplained exceptions can "disprove" an entire, widely accepted theory. But even though to a scientist like myself this logic smacks of desperation, these people have faith on their side, and there are many who are prepared to listen.

WHAT ABOUT HUMAN MORALITY?

Initially, an apprehensive Darwin chose not to write about the greatest controversy of all: the application of his new theory to human beings. But when *The Descent of Man* was finally published in 1871, the scenario

Darwin put together was nothing short of remarkable. Not only did he at least outline what he could of the human origin story in the context of an evolutionary sequence starting with apes, but also in certain areas he even managed to provide important environmental details and specify some likely selection mechanisms. For human physical evolution, particularly of our outsized brains and upright locomotion, Darwin's hypotheses were daring and, given the dearth of information in his day, keenly prescient. The basic outlines he set down on paper are still valid today.

Another of Darwin's equally daring hypotheses had to do with the origination of moral behavior and the human *conscience*—the subject of this book. His treatment of a self-conscious conscience was particularly provocative because now he was bringing his naturalistic approach close to the soul, previously the exclusive purview of the church or, more precisely, of God. Darwin did not take on the problem of how human beings came to have a soul; indeed, the word does not even appear in the lengthy and detailed index for *The Descent of Man.* But he clearly thought our conscience and moral sense were as "naturally selected" as our large brains, our upright posture, and our general capacity for culture.

As a brilliant and meticulous scientist, Darwin didn't have the data to make anything like a plausibly scientific case with respect to conscience origins, but he did the best he could, and under the circumstances his best was quite good. In an 1871 passage concerned with instincts of "sympathy," which is often quoted by contemporary scientists such as evolutionary biologist Jessica Flack and primatologist Frans de Waal and by others who take an interest in moral origins,[11] Darwin wrote, "Any animal whatever, endowed with well-marked social instincts, the parental and filial affections being here included, would inevitably acquire a moral sense or conscience as soon as its intellectual powers had become as well developed, or nearly as well developed, as in man."[12]

An introspective Charles Darwin waxed eloquent on how the conscience worked, and his own superego was obviously strong and active. Socially, it was this charitable "inner voice" that kept us from getting in trouble with our fellows, Darwin told us, and he wanted badly to explain its evolutionary origin. But all he could tell his readers was that gaining a conscience and hence a sense of morality was, in effect, an inevitable outcome if a species became sufficiently smart and socially sympathetic to reach the human level.

Unfortunately, this gave our uniquely human conscience the evolutionary appearance of being a mere byproduct, a side effect of intelligence and sympathy. This is a position I think we can vastly improve upon with present knowledge, and in the chapters to follow I will bring in some quite specific hypotheses to explain how the conscience evolved and why it did.

THE MYSTERY OF HUMAN GENEROSITY

Darwin tried to answer yet another profound question: Why do human degrees of generosity seem to defy the patently "selfish" principles of natural selection theory? This original puzzle was influentially redefined in modern terms in the 1970s by scholars such as social psychologist Donald T. Campbell and biologists Richard D. Alexander and Edward O. Wilson.[13] And for more than three decades now, a major and growing interdisciplinary academic industry has devoted its efforts to resolving the "altruism paradox"—with only partial success. I hope that this book will get us closer to a scientifically satisfying answer.

The historical background for this fascination with altruism is quite interesting. Darwin's "selfish" theory of natural selection held that individuals were indirectly competing for fitness and that, as we've seen, those who were more vigorous or otherwise better adapted for securing food or mates would come out ahead in shaping the hereditary future of their species. This could be explained simply: favored individuals had more surviving and breeding offspring than others in their group or region or wider population. But Darwin also realized that this kind of advantage could be helped along by family connections because close relatives who naturally helped each other tended to share the same heredity. He didn't carry this second idea very far, but it turned out to be extremely important.

This became clear a century later when the well-known population geneticist William Hamilton showed through mathematical modeling that selfishly competing individuals could make reasonable personal sacrifices if these benefited their own offspring.[14] Because, on average, an individual shares 50 percent of his or her heredity with progeny, investing in offspring helps propagate the genes that individual carries. The same 50 percent rule holds for helping siblings or parents. Being generous to a grandchild or a first cousin (a relationship that involves only a 25 percent genetic similarity) still makes sense if the costs of helping aren't too high and the benefits are significant. This powerful theory, known as kin selection, can apply to even lesser degrees of kinship as well, as long as the donor's costs are sufficiently modest and the benefits sufficiently large.

Darwin identified a further problem that continues to perplex scholars today.[15] In real life, humans don't merely assist their close or distant blood kin; they also help people who are unrelated to them—even though from a biological standpoint giving such altruistic assistance will be costly to their fitness because there is no shared heredity. This means that unless these unrelated beneficiaries somehow are reciprocating in something like equal measure, or unless some other type of "compensation" exists, the individuals who take such action may be lowering their own fitness and raising the fitness of their partner. At its simplest, the evolutionary lesson is all too obvious: in theory, generosity should stay within the family because nepotists who refrain from altruism will be able to outcompete the altruists.

Another modern biologist, George Williams, provided another big reason that mutant genes that made for generosity outside the family wouldn't stick around for very long.[16] Free-riding genes—genes that send their bearers the opportunistic behavioral message "take from altruists but avoid giving"—should increase in frequency as they replace the genes of the altruistic "losers." Still scratching their heads in many ways about this matter of our remarkable human generosity are scores or, more properly, thousands of evolutionary biologists, ethologists, anthropologists, sociologists, and philosophers, not to mention a wide array of evolutionary psychologists and a significant cohort of evolutionary economists. All continue to work on problems germane to the basic evolutionary puzzle of altruism and the closely associated problem of "free riders," which together have been a major interest of the aforementioned academic industry for almost four decades. At this writing, at best the evolution of human generosity is only partly explained.[17]

THE MYSTERY OF EXTRAFAMILIAL GENEROSITY

Unfortunately, in the field of evolutionary biology "altruism" has become a technical term that, after almost half a century of intense debate, still remains to be used with total consistency.[18] Sometimes, for instance, it means being genetically generous to anybody at all, including kin; sometimes it means being generous to people lacking any blood ties to the generous party. Because of the focus of this book, I will be relying on the latter meaning, using "altruism" and "extrafamilial generosity" as exact synonyms, while when costly help is given to relatives I will call this "nepotism." The altruistic type of beneficence may refer either to acts of costly generosity toward specific unrelated individuals or to sacrifices of personal interests as the individual contributes to enterprises that benefit the community as a whole. Thus, altruism and the possibilities for human cooperation are intertwined, for altruistically generous individuals make for superior cooperation in groups that include nonkin.

In biological terms, then, when we speak of altruism, we're speaking of behavioral tendencies that dispose people to give more than they receive in terms of acts that reduce relative fitness.[19] Even if all of the underlying genetic selection explanations are not yet fully developed, the tangible behaviors are obvious enough. People predictably open their veins to anonymously give blood or open their wallets to help starving children in developing countries, and generous assistance following a natural disaster anywhere on the planet can be quite impressive. Therein lies the altruism puzzle: Why do so many members of a supposedly egoistic and nepotistic species in some contexts become quite giving to people they aren't related to and sometimes don't even know?

HOW "GOLDEN RULES" AMPLIFY OUR INNATE GENEROSITY

Even though common sense alone can tell us that our dispositions to extrafamilial generosity are significant, it's equally obvious that this is rather negligible compared to our truly powerful dispositions to egoism and to nepotism. It also is apparent that these genetic dispositions don't *determine* our actions. Rather, they set up the behaviors in question so that they will be exceptionally easy to learn.[20] Thus, we must consider the interaction of genes and culture, and the influence of the social environment on how we behave should not be underestimated.[21] For instance, actively preaching a "Do unto Others" ideology can powerfully reinforce the relatively modest innate tendencies that favor extrafamilial generosity and thereby enable groups to work better together.[22] I will discuss my research findings in this area in Chapter 7, as well as in Chapter 12 toward the end of the book.

When individuals in a nomadic, egalitarian hunting band seek to promote generosity, they recognize that self and family will always come first and that therefore people will need some special "persuasion" to contribute robustly to the group as a whole. In short, band members understand that if they are to better reap the benefits of group cooperation, they'll need to apply their local version of the Golden Rule manipulatively, as a refined type of social pressure designed to bring out the best in human nature.[23]

Keep in mind three things about classical hunter-gatherer bands. First, they always involve a mix of related and unrelated families.[24] Second, they predictably cooperate in certain activities with no expectation of immediate or exact reciprocation.[25] And third, they actively preach in favor of wider generosity within the group, precisely because human propensities to be selfish or nepotistic are so strong in our species.

In my research,[26] I have found such strictures in favor of extrafamilial generosity to be both prominent and probably universal in these mobile band-level cultures—whose lifestyles are similar to those of the prehistoric foragers who basically had evolved our modern set of genes for us by 45,000 years ago. It would appear, then, that these strictures are fairly ancient. The people involved obviously know what they're doing when they apply such pressure, and as an anthropologist I fully agree with the everyday intuitions of these socially manipulative hunter-gatherers. I, too, believe that our relatively modest propensities to engage in extrafamilial helping behavior can provide an important basis for human cooperation—and do so all the better if such propensities are being strengthened by prosocial socialization of children, by the application of positive social pressure on adults to behave with generosity, and by the discouragement (or elimination) of selfish bullies and cheaters, who hamper cooperation and also create conflict.

Of course, I have the anthropological advantage of knowing that in later, larger types of societies such as chiefdoms or early states, the same prosocial propensities can contribute to still greater community cooperation—and that today, as with both a ruthless Nazi Germany and Great Britain as Hitler's wartime adversary, cohesive cooperativeness can at least approach the truly selfless, "eusocial" collaboration that takes place in anthills. In the case of these insects, however, the cooperating individuals tend to be close genetic relatives, so their apparently "selfless" contributions to group interests can be explained nepotistically by kin selection, combined with group selection.[27] It is with the genetically "reckless" generosity of humans where generosity extends beyond nepotism to nonkin that the major evolutionary puzzle arises: How can such natural dispositions stay in place, especially if opportunistic free riders are poised to outcompete those who are extrafamilially generous?

CAN GROUP SELECTION SOLVE THE PUZZLE?

Darwin saw this problem very clearly, even though he couldn't begin to fully anticipate the sophisticated modern population-genetics models that began to emerge in the 1930s, which resulted in systematic theories such as those of Hamilton and Williams, and eventually led to Edward O. Wilson's global redefinition of human social biology in terms of the altruism paradox we've been discussing. A century earlier Darwin simply wondered how he could ever reconcile his new theory, which was so "individualistic," with the fact that patriotic young men so willingly went to war to sacrifice their lives for their countrymen. They were sacrificing not only their lives but also the lives of their future progeny, who otherwise would be inheriting these generous tendencies. The great naturalist was confounded.

Darwin had in mind the fact that free-riding cowards would be avoiding these same risks and that their greater numbers of surviving offspring would be inheriting the same selfish tendencies. In short, following his theories, generously self-sacrificial patriotism should always be on the wane, while dispositions to hold back and stay safe should always be proliferating. This meant that over the long run any tendencies to sacrifice personal interests for the good of the group should be automatically suppressed by natural selection—yet in practice young men were going to war, and many were doing so eagerly.

Darwin did offer a possible solution to this puzzle. Here are his famous, often-quoted, somewhat convoluted words, taken from *The Descent of Man:*

> It must not be forgotten that although a high standard of morality gives but a slight or no advantage to each individual man and his children over the other men of the same tribe, yet that an advancement in the standard of morality and an increase in the number of well-endowed men will certainly give an immense

> advantage to one tribe over another. There can be no doubt that a tribe including many members who, from possessing in a high degree the spirit of patriotism, fidelity, obedience, courage, and sympathy, were always ready to give aid to each other and to sacrifice themselves for the common good, would be victorious over other tribes; and this would be natural selection.[28]

This brilliant piece of reasoning still haunts the large community of scholars who study human social evolution. Group selection theory was for a long time spurned by the great majority of biologists, although today it has found its place in multilevel selection approaches.[29] E. O. Wilson led the earlier charge against naïve group selection theories, but today one explanation for either nepotism or altruism is that groups having more or better cooperators will outreproduce lesser groups. This level of explanation will not be prominent in the pages that follow, because the emphasis will be placed on collective punishment and free-rider suppression as these affect selection taking place between individuals *within* groups.

THE WORLD'S FIRST CROSS-CULTURAL RESEARCH

I'm confident in saying that Darwin strongly desired an explanation of conscience and morality in a full natural-history context, which would make explicit how such remarkable human capacities could have developed over time. This would have required him to specify the type of favorable environmental conditions that would have prevailed and which mechanisms of selection could have contributed to this historical evolutionary process of moral origins. But Darwin was not able to accomplish this, not because of any lack of insight or ambition, but because in his time he lacked the necessary data from primatology, paleoanthropology, cultural anthropology, and psychology, along with explanations of brain functions from cognitive neuroscience. All of these fields have either emerged since Darwin's time or else have grown by leaps and bounds so that today we may finally have the scientific information needed to put together a plausible evolutionary scenario.

Darwin was not given to being reckless in reaching scientific conclusions, but we might ask why he didn't at least speculate about the possible specifics of conscience origins. There are several answers, probably. First, the archaeological record of his time was woefully inadequate, comprising only a scattering of fossilized bones and a few stone tools our predecessors had left behind. Second, very little was known of brain functions relevant to our sense of right and wrong or of how African great apes—as potential "stand-ins" for our distant ancestors—behaved outside of zoos. Third, the science of ethnography was too nascent to systematically look for universals in social behavior that could then be tied to our biological nature.

What Darwin did about this last problem was quite remarkable. He initiated the first systematic research across cultures by writing to colonial administrators and missionaries all over the world to ask them whether indigenous people in Asia, Africa, and elsewhere blushed with shame.[30] Having one's face "color" for social reasons is unique to humans, and Darwin was interested in knowing whether morally based, shameful blushing was merely something that certain groups did because their local cultures led them in that direction or whether, as he suspected, there might well be a strong hereditary component. What his far-flung anthropological research project told him was that indigenous people everywhere did seem to blush with shame. And on this basis he could assume that, as an important aspect of our conscientious moral sense, human shame reactions surely had to have an innate basis.

This research project stands today as a true landmark in the anthropological science of human nature, and what it suggested more generally was that conscience and morality had to have *evolved,* in the biological sense of the word. Carrying this line of research forward, I shall show that the human conscience is no mere evolutionary side effect, as Darwin had to imply it was. Rather, it evolved for specific reasons having to do with the Pleistocene environments humans had to cope with prehistorically and, more specifically, with their growing ability to use group punishment to better their own social and subsistence lives and create more socially equalized societies.

SOCIAL SELECTION AS "PURPOSEFUL" NATURAL SELECTION

There are several ways that social preferences of humans can affect genetic outcomes.[31] One is that as individuals people may choose others with good reputations as marriage partners or as partners in cooperation, which helps their fitness. The other is that entire groups may come down hard on disliked social deviants, which damages their fitness.[32] My general evolutionary hypothesis will be that morality began with having a conscience and that conscience evolution began with systematic but initially *nonmoralistic* social control by groups. This involved punishment of individual "deviants" by angry bands of well-armed large-game hunters, and like the preaching in favor of generosity that followed, such punishment could be called "social selection" because the social preferences of group members and of groups as a whole were having systematic effects on gene pools.[33]

The punishing of deviants occurs because people feel individually threatened or dispossessed by social predators, but also, in a larger sense, because socially disruptive wrongdoers so obviously lessen a group's ability to flourish through cooperation. Thus, this punitive side of social selection involves at least an immediate kind of "purpose" in the sense of large-brained humans actively and often quite insightfully seeking positive social goals or averting social disasters that can grow out of conflict. It's no surprise that the genetic consequences, though unintended, go in the direction of fewer tendencies that make for social predation and more tendencies that make for social cooperation. Therefore, on an everyday basis group punishment can improve the immediate quality of group social life, even as over the generations it can gradually shape the genotype in similar directions.

That group members' punitive actions can not only influence group life but also shape gene pools in similar directions is one major thesis of this book. Therefore, we must ask if some limited purposeful element is actually creeping into a biological evolutionary process that, in theory, is supposed to be operating "blindly." That is, could social selection be introducing what might be called some "lower-level teleology" in the sense that some purposeful inputs could be influencing natural selection process?[34] Such a theory modifies somewhat one of the most basic premises of modern Darwinism: that natural selection simply organizes itself, merely a*ppears* to be "solving problems," and basically is *blind*.[35] Thus, in the words of biologist Ernst Mayr, Darwinian selection is "teleonomic," rather than teleological.[36]

Mayr was referring to *natural selection* as a basic overall process. Of course, two totally unambiguous and potent practical examples of *purposeful* selection would be animal breeders and modern genetic engineers. We must also include members of discredited eugenics movements, for the Nazis knew exactly what they were trying to accomplish. All three consciously want to tamper with gene pools, and they all have some insight into what they are attempting.

It's with good reason that we don't think of prehistoric hunter-gatherers as these kinds of active agents at all. Yet I shall propose that *unwittingly* their social intentions did affect gene pools in ways that were predictable, highly significant, and at least were guided by rather sophisticated immediate purposes that had to do with improving their quality of life. Prehistorically, I believe that this provided a special "focus" to the process of human social selection, a focus that derived from the very consistent practical purposes of the actors. They were moved to persuade people to behave more altruistically and also to deter the free riders in their midst, and both affected not only their immediate everyday life but also their gene pools long term.

A "NEW" WAY OF USING DARWIN

I follow Darwin in thinking that the analysis of evolutionary developments over time can produce powerful explanations, especially if it includes abundant naturalistic detail. However, such holistic natural-historical approaches appear to be old-fashioned, for nowadays evolutionary research usually is done piecemeal, attacking one delimited problem at a time, and the modeling of behavior and its effects on gene pools is approached logically, in terms of "design" and "adaptation." What has been set aside very often is an actual Darwinian analysis that focuses on the historical dimension.

In looking at the effects of social selection over scores of millennia, I will be developing a rather novel evolutionary scenario by today's standards. My idea will be that prehistorically humans began to make use of social control so intensively that individuals who were better at inhibiting their own antisocial tendencies, either through fear of punishment or through absorbing and identifying with their group's rules, gained superior fitness. By learning to internalize rules, humankind acquired a conscience, and initially this stemmed from the punitive type of social selection I mentioned previously, which also had the effect of strongly suppressing free riders. Later, I shall argue that a newly moralistic type of free-rider suppression also helped us evolve our quite remarkable capacity for extrafamilial generosity.

The next several chapters will concern the evolutionary background for these moral origins, including a realistic discussion of whether certain other animals may be on the road to morality, and a detailed description of the social behaviors of our very distant ancestors, who of course were apes, just as Darwin told us. In Chapter 4 I will also be reconstructing the behavior of the first fully "modern" humans, as of 45,000 years ago, for they are basically the end point for moral evolution in the biological sense. Today, even though we live in cities and write and read books about morality, our actual morals are little more than a continuation of theirs.

The Darwinian evolutionary analysis proper will begin in Chapter 6, initially focusing on moral origins in a natural Garden of Eden and more specifically on the conscience and how this uniquely self-conscious agency came into being as a result of a punitive social environment.[37] This development has profound adaptive implications for our species, both prehistorically, when it made large-game hunting a more viable and useful enterprise for our predecessors, and today, when we remain moral and continue to benefit from being moral.

If we think of a society of modern people with no conscience or sense of right and wrong, it becomes difficult to imagine existing in present-day large, anonymous urban environments, where crimes against both society and individuals are so difficult to detect. That most people have strong and active consciences benefits all of us; at least potential wrongdoers can't hide from their own consciences even if these settings all but invite them to act as social predators.

Earlier, this moralistic type of consciousness helped culturally modern hunter-gatherers to navigate their courses socially in small intimate bands, where police detectives weren't needed because social deviants were so readily identified, and controlled, by their gossiping peers. In such bands the fact that people had well-developed consciences enhanced group social life because this inner voice slowed down the antisocial deviant tendencies that individuals in these groups harbored, and thereby helped to reduce conflict within the group and make cooperation more viable.

My modern-day emulation of Darwin's historical type of analysis in the chapters to come will be both somewhat novel and, I hope, plausible. And once the important question of conscience origins is resolved, we'll be in a far better position to explain how humans acquired the unusual (and to some, all but inexplicable) degrees of sociality and sympathetic generosity that allow us to cooperate as willingly as we do. As will be seen, if we had never gained some kind of a conscience, which gave us a primitive sense of right and wrong, we would never have evolved the remarkable degree of "empathy" and the accompanying traits of extrafamilial generosity that enrich human social life as we know it today.[38]

2 Living the Virtuous Life

WHICH IS UNIVERSAL: SHAME OR GUILT?

There are two ways of trying to create a good life. One is by punishing evil, and the other is by actively promoting virtue. My evolutionary theory is that the punishment of deviant behavior is older, so several chapters in this book will be devoted to crime and punishment and their deep evolutionary background. Later, in Chapter 7, I will take up the positive side of social interactions, which includes preaching in favor of being generous to others and of being helpful even if the recipients are not members of a person's own extended family and are not necessarily helping back.

To begin this discussion, we'll need to consider the human emotions that make for a sense of right and wrong.[1] An evolutionary view needs to be global, and to achieve this, ethnocentrism must be avoided. Guilt is frequently on the tongues of Americans and for that matter on the tongues of Christians and Jews everywhere, but it is not very much on those of Buddhists or Hindus or of Confucians or followers of Islam. Although the word is not easy to define and definitions may vary, for most people "guilt" seems to mean an inward, private focus arising from the experience of negative feelings about past misdeeds or sins. "Shame" has more the meaning that a past malfeasance has become known to others, or might well become public. Although either guilt or shame may lead to remorse, shame, with its more outward focus, seems to be more salient for many people raised in Asia, where the idea of "face" is important, or in the Middle East, where "honor cultures" are so prominent. The Garden of Eden fable tells us that shame also is important for Christians and Jews insofar as the Middle-Eastern-based Old Testament has trained us to think in terms of fig leaves and blushing—with shame.

To avoid confusion, and also to avoid placing my own Western perspective at the fore, I'm going to make a choice and use "shame," rather than "guilt," when speaking of uneasy or painful feelings people have about their present or past morally reprehensible deeds. I've made this simplifying decision not because Asians and Middle Easterners outnumber guilt-ridden Judeo-Christian Europeans and Americans, but for a quite different set of reasons that strike me as being anthropologically sound.

First, a moral word similar to "guilt" is not to be found in many world languages, including those of hunter-gatherers and tribal people. However, "shame words" do appear everywhere,[2] and they seem to be quite prominent in people's minds. Furthermore, shame feelings are directly linked to a universal human physiological response that is triggered by a sense of moral culpritude—*blushing*—whereas guilt has no such physical correlate as far as we know. When Darwin saw this connection between shame and blushing and deemed it important, he was right. Shame will be a key universal concept here as we begin to deal with the evolutionary basis of the human conscience.

DO DOMESTICATED DOGS HAVE A SENSE OF SHAME?

In thinking about moral behavior, Darwin opened his mind to ask if other animals might also have a sense of right and wrong.[3] After devoting considerable thought to the matter, I've come to the personal conclusion that, although chimpanzees and for that matter domestic dogs are very good as learners of rules, humans may be the only animal species to deal moralistically in virtue and evil and to internalize rules on that basis. If any other animal had such abilities, most likely it would be a highly social animal like an African great ape or perhaps a socially sensitive carnivore like a wolf or a dolphin.

I'm quite certain that many people with beloved pets would disagree, dog owners in particular. Many sense that their animal companions are feeling *morally* chastised when told, "You should be ashamed of yourself," just as they may seem to respond proudly and virtuously to "That's a *very* good dog!" I've experienced such anthropomorphic reactions myself, with delight born out of a sense of kinship, but obviously that doesn't make such a reaction scientifically the case.

Darwin focused on dogs because they're unusually congenial to us psychologically and because dog owners have such a well of experience, so many tales to tell about their humanlike pets. In fact, he gathered a large corpus of stories that were suggestive of canine sympathy, loyalty, and self-sacrificial protectiveness, along with a few anecdotes that might have suggested the presence of guilt feelings or shame. But this open-minded scientist did not jump to conclusions.

It is with a wistful sense of personal regret that I must inform other loyal dog owners that when their charges *seem* to be giving them guilty looks, in all probability they (the owners) are projecting their own moralistic human reactions onto an amoral canine. An empathetic dog may be feeling uncomfortable in the face of disapproval, or submissively fearful of punishment for breaking a rule, and it may be showing this eloquently by means of body language, but I'm reasonably certain that a humanlike sense of being *ashamed*—and I mean feeling shame because there exists a strong and moralistic emotional identification with a serious and important rule that has been broken—plays no part in this picture.

Charitable interpretations with respect to doggie shame or doggie guilt are scarcely surprising, precisely because we humans have been breeding dogs to have feelings similar to our own for at least fifteen thousand years. Today this is done very methodically, but in the distant past simply favoring puppies that made the best pets, and then doing so over many generations, would have modified the "basic personality" of domesticated canines.[4]

As a fanatical dog lover, I would be the first to say that the dogs we've domesticated truly are friendly, affectionate, loyal, empathetic, eager for approval, and, if their masters are in trouble, often protective and self-sacrificing. If properly trained, they are as good as we are at following rules, and with all of these similarities it is natural to expect them to also have feelings of shame. But moral they are not, for a rule-internalizing conscience and sense of shame would appear to be missing. I realize that my skepticism is a matter of opinion and that a human can never get inside the head of a dog. However, there are at least a few facts that tend to support this hard-nosed viewpoint.

It's easy enough to think you're seeing a dog conscience in action when you come home to find not only a mess on the floor but also a cowering canine with head bowed, ears back, and tail between its legs. It also seems logical that if you then punish this humanlike culprit in the presence of its misdeed, it will recognize the shameful error of its past ways and desist in the future, just as a human would, because shame feelings are unpleasant and are to be avoided. And it's certainly true that nasty nose rubs or training whacks with rolled-up newspapers will be remembered by your dog—as evidence of a beloved master's obvious disapproval. In that sense dogs can learn our kind of rules, for we've bred them to be sensitive this way for thousands and thousands of generations by favoring the more docile individuals.

However, the idea that *after-the-fact* punishment can produce a positive shift in the dog's behavior, just as it does with humans, is quite erroneous. Any professional dog trainer will tell you that you must punish your canine pet right in the commission of the deviant act—or at most within just *six-tenths of a second* after the dog's unappreciated deed is done.[5] Otherwise, apparently your dog will be confused because it will see you, a person it is closely bonded to, being hostile or hurting it for no good reason. People, on the other hand, understand perfectly well when they are punished now for a previous rule infraction, and as we'll be seeing in Chapter 5, so can an African great ape. But in this respect dogs seem to live only in the present.

A devoted dog owner could argue that nevertheless dogs *must* be feeling shame—just look at their body language and those eyes. I can't prove that to be false. What I can point out objectively, however, is that dogs neither blush with shame as we do nor seem to respond to punishment after the fact. Thus, in spite of being selected for humanlike qualities during the course of many thousands of dog generations, dogs remain on a significantly different wavelength regarding ex post facto condemnation and punishment.

"Might makes right" is what prevails among the ancestors of all dogs, the wolves. In every pack there are alphas who impose their rules of dominance upon subordinates, and if a subordinate successfully breaks a rule—when it gets away with something behind the alpha's back—there's absolutely no evidence of "shame" or "remorse" in the sneaky, willful subordinate's body language. A subordinate caught in such an act will certainly try to appease the superior, but this has nothing to do with feeling *morally* reprehensible. It's simply a matter of manipulative self-protection, and this, too, is found in humans. The difference is that we're also moral.

It appears that the minds of dogs continue to be genetically set up to make them respond to punishment in only a very immediate way and that for some reason, yet to be discovered, this particular piece of brain wiring has resisted the attempts of egoistic humans to modify their domestic dogs and make them into obedient companions who totally remind us of ourselves. One major hint is that in dogs the prefrontal cortex—the part of the brain that helps in making social decisions that result in self-control—is much smaller, proportionately, than it is in human brains. Perhaps the potential just wasn't there, even though as dog breeders we humans have tried hard enough to make our pets as prosocially oriented as we are.

MORALLY DAMAGED MINDS

Some of the most interesting things we know about our brains and their connection with morals come from what we can learn about a very small proportion of our human populations—people who most decidedly, to judge from their attitudes and behavior, are quite "amoral." Although many seem to be born that way, a few others have suffered brain traumas with remarkable and revealing effects. If early in life the prefrontal cortex (residing just behind the forehead) of healthy, "normal" children is physically damaged, they may then grow up without the ability to understand and follow rules or to deal with authorities. Because their sense of right and wrong is impaired, they may find it difficult or even impossible to plan reasonably successful social lives.

Neuropsychologist Antonio Damasio reports on several cases. A year-and-a-half-old child had her head run over by a car, with no ostensible negative effects after a few days.[6] Her behavioral problems came to light only when as a three-year-old she proved to be unusually unresponsive to parental rules. (I should mention that, morally speaking and otherwise, her parents were perfectly normal people.) Subsequently, she grew up to be an impulsive petty thief who simply couldn't follow rules well enough to keep a job, showed a pathetic lack of empathy for her own baby, and didn't really seem to understand the difference between right and wrong. Unable to evaluate and control her own impulses, she couldn't function properly as a social being, and her life was a mess.

Just as this woman's brain showed damage to the prefrontal cortex, so, too, did the brain of Phineas Gage, a responsible and amicable late-nineteenth-century railroad worker who became famous in the annals of psychology. Phineas was involved in an accident in which a metal spike went up into his eye socket and out through the front of his head, damaging this same brain region. Gage was able to stand up and think and speak immediately after the accident, but his personality was changed on a permanent basis. He lost his affability and became impulsively irritable, obscene, and impossible to get along with. Tragically, Phineas could no longer hold a regular job, and he ended up as a circus sideshow freak.[7]

An equally telling case was that of a schoolteacher, happily married at forty, who was caught by his utterly surprised wife looking at Internet pornography involving children and subsequently tried to "come on" to an eleven-year-old girl. His lack of impulse control led to divorce and possible prison. Eventually, the poor fellow was diagnosed with a benign tumor that was pressing on his prefrontal cortex—and when the tumor was removed, he went back to being normal. When the tumor reappeared, his deviant interests again became uncontrollable, so the cause and effect relationship was all too clear.[8] As a brain area devoted in general to planning, the prefrontal cortex helps us to assess social consequences and also to control antisocial impulses. These functions go far in defining the human conscience as a faculty that enhances our personal fitness by keeping us out of trouble with our groups.

THE BRAINS OF PSYCHOPATHS

Then there are those who are *born* "impaired." Psychologist Robert Hare spent his life studying criminals he objectively evaluated as being "psychopathic." One aim of his screening test—he was the first to develop one that cunning psychopaths couldn't outwit—was to identify such people in prison and keep them off the streets.[9] Hare's assumption was that psychopaths had inherited characteristics that kept them from developing normal consciences based on the usual *moral emotions,* which include a deeply felt sense of right and wrong and feelings of empathy for others. These unusual people range from deadly and unrepentant serial killers, some of whom we all know by name but many of whom we never manage to identify or police never manage to catch, to a much larger number of often glib and sometimes quite charming but utterly egoistic and unempathetic con artists, who lie recklessly and predictably exploit and harm other people without remorse or shame.

Whether psychopaths devote themselves to murder or to street crime or white-collar crime or con games that exploit the gullible, they are unusually given to domination or control, and what they all share are a lack of a normal moral compass and little trace of concern for the damage they are inflicting on the trusting souls they exploit. In lacking a normal conscience that includes making emotional connections with others, they lie without compunction in order to selfishly exploit others—and they fail to feel any sympathy for those they defraud or murder. These people are more frequently male than female, and in general their emotions are shallow, with a lack of the feelings that connect ordinary people with the moral rules they are able to internalize as children. Psychopathy shows up early in life, so in an important, emotional sense the learning of moral rules is incomplete.

I should hasten to emphasize that the typical psychopath is not the popularly conceived serial killer but is simply a con man who has no empathy for those he despoils. Typically, he's intelligent, self-centered, and good at putting on a convincing face, even though sometimes his lying becomes obvious. He's the perfect candidate to be a seller of bogus stocks to retirees or to be a battering husband whose wife doesn't realize he was born that way and keeps on hoping he'll reform. However, if killing does attract him, he kills without mercy, and the psychopath's hall of fame includes the Hillside stranglers—there were two of them, cousins, who mercilessly tortured their California victims—along with, of course, John Wayne Gacy and Jeffrey Dahmer. *Silence of the Lambs* aside, those are the ones we've heard of, but there obviously are many more of them, mostly uncaught, who thrive in modern urban societies, where a cloak of anonymity is available to the nefarious. With no very active conscience functions, and often with no fear of God's punishment, they only have to stay ahead of the police.

Psychopaths are absolute masters at dissembling, and they understand how moral feelings operate, even though personally they live in an emotional wasteland. Curiously, this enables them to be true experts at manipulating the feelings of others, and that's why they've so often had their way with parole boards, convincing them that if only they are given the opportunity, they'll reform and become contributing members of society. Remember, psychopaths are born that way, so the normal hookup between positive emotions and a sense of "rules" cannot develop. That's why Hare's book is titled *Without Conscience*. The best we can expect of a psychopath, Hare tells us, is that as old age approaches, his (or her) antisocial tendencies will tend to dwindle somewhat on their own. Otherwise, parole boards beware.

Typical psychopaths have suffered no trauma to their prefrontal cortexes, so they are natural-born immoralists. Because they tend to have above-average intelligence, some of them are unusually adept at understanding societal rules, and they can be brilliant in pretending to have normal emotions. In addition, they can be quite astute in exploiting the trust of normal people: as one example, a serial killer lured a series of well-meaning, doomed girls to his car by asking for assistance with his "broken leg." However, many less careful psychopaths tend to be reckless thrill-seekers, who spend most of their lives behind bars as relatively petty criminals who predictably get caught.

People very significantly afflicted in this way probably number as high as one or more out of several hundred in our total population. Not only are they without feelings of remorse or shame; often, they also take a curious kind of pride in their impulsive depredations. These individuals crop up in every walk of life, and the disorder is so fundamental that if they rate high on Robert Hare's screening test, psychiatric treatment does them little or no good. They will never be fully moral for two reasons: first, they haven't these emotional hookups, which are needed to identify with and internalize society's rules, and second, they lack empathetic concern for others.

HOW NORMAL PEOPLE "CONNECT" WITH GROUP MORES

Half a century ago, sociologist Talcott Parsons tried to look at human social behavior from a number of perspectives, including the cultural and the psychological, and he spoke convincingly of the "internalization" of values and rules.[10] By this he meant, in effect, that when groups translate their social values into rules of conduct, such as "Do unto Others," individuals form emotional connections with those rules so that they feel fine about following them and uneasy about breaking them. Going against an internalized rule such as "Thou shalt not steal" can make normal people—but not psychopaths—feel apprehensively ashamed of themselves, socially tortured if they are caught in the act, and sincerely remorseful after the fact.

Let's consider Parsons's insight as it relates to psychopaths. These predators can't *internalize* society's values and rules as normal people do, and as a result they lack the active "inner voice," laced with self-judgmental moral feeling, that Darwin talked about so eloquently.[11] In contrast, the rest of us will feel that an important part of our identity is tied in with how we follow rules, and our self-respect will suffer or prosper accordingly. Morally normal human beings identify strongly with their own cultures and with the specific rules inherent in living a productive social life. They do so even though their appetites for power or "things" or sex or status may easily lead them to break some of these rules. Psychopaths simply don't identify with the rules in the first place.

How can such a pathology evolve? How is it that at least one person in perhaps a few hundred can be a seriously predatory psychopath today? How can these socially perverse people's genes stay alive and well in the human gene pool given that seriously deviant types are punished severely, with consequences for their fitness? To answer this question, we must rephrase the problem in evolutionary terms and ask, what benefits could psychopathy have brought to individuals who carried this trait in our hunter-gatherer past?

Prehistorically, some of these people surely became targets of capital punishment, but perhaps there were also some fitness advantages. For instance, that such individuals were unusually selfish with tendencies to dominate might have worked for them quite well in the strongly hierarchical early human communities that predated an egalitarian human lifestyle based on systematic group punishment.

And even in nonhierarchical prehistoric band societies that were moralistically egalitarian, being unusually "selfish" could have had some fitness payoffs—even though if expressed in *seriously* antisocial ways, it could have led to big trouble. As we'll see, the name of the game was self-control.

It seems likely that there would be some specific genes involved, but how would this be demonstrated? Like Robert Hare, psychologist Kent Kiehl, whom Hare mentored, works with prison populations. He's an innovator: he brought mobile MRI units right into prison yards to scan the brains of felons. This research professor at the University of New Mexico's Mind Institute used Hare's formal evaluation procedures to decide reliably which criminals were psychopaths and which ones had a normal, "emotionally connected" sense of morality, and then he compared the two.[12]

A murderer could fall into either category, depending on whether a moralistic sense of empathy with the victim led to feelings of remorse afterward. Whereas typically a psychopathic murderer felt untroubled by the killing during and afterward, a morally normal felon who killed someone in a flash of uncontrollable anger would be deeply upset with himself over the ultimate hurt and damage he had done. This demonstrates that his moral makeup was no different from anyone else's, and the remorse could be lifelong. After doing brain scans on large samples of prisoners in both categories, Kiehl noted that there were apparent anomalies in the paralimbic systems (at the base of the brains) of the psychopaths. This fairly old component of the brain facilitates the coupling of emotions with how people react to a variety of social situations, so this brings us back to the matter of rule internalization as a normal function of the conscience.

THE BEST-ADAPTED CONSCIENCE IS FLEXIBLE

Having a conscience can be considered in a number of ways. For instance, Sigmund Freud talked about the superego as a mechanism of the mind that stands between us and our unruly libidos.[13] And economist Robert Frank has made the case that a conscience, with its emotions, is individually adaptive.[14] In more general usage, having a conscience simply means being internally constrained from antisocial behavior, and, I would add, deriving one's self-regard from following society's rules. Here, however, the evolutionary conscience will be defined still more broadly.

Decades ago, in *Darwinism and Human Affairs,* biologist Richard D. Alexander defined the evolutionary conscience as being more than an inhibitor of antisocial behavior. He called it the "still small voice that tells us how far we can go in serving our own interests without incurring intolerable risks."[15] Thus, a conscience seems to be as much a Machiavellian risk calculator as a "pure" moral force that maximizes prosocial behavior and minimizes deviance. If we are interested in the conscience and its evolution, we must define it dispassionately in terms of how it has served our fitness, and in this respect Alexander's realistic definition is a bit better than Darwin's. Of course, Darwin saw the conscience as a means to inhibit immorality, rather than to strategize how much immorality a person might get away with. Simple introspection will tell us, if we are honest with ourselves, that an evolutionary conscience does both.

In this context, we may ask exactly how strong the emotions are that bond *normal* human beings to their group's rules. Internalization doesn't mean that our best citizens become so deeply involved with society's rules that they follow them automatically without thinking about alternatives—especially if socially disapproved alternatives happen to offer great satisfaction. Not at all. As we all know from personal experience, the selfish needs and desires that orient our behavior and generally help our fitness provide many social temptations that can result in moral censure and even personal disaster. What well-internalized moral values and rules do is to slow us down sufficiently that we are able, to a considerable extent, to pick and choose which behaviors we care to exhibit before our peers. As a result, most of our self-interested acts don't become so predatory or antisocial that we're likely to be discovered and severely punished—with our fitness ultimately being damaged.

Thus, our consciences can often make us into ambivalent conformists when an attractive but socially disapproved behavior presents itself. Imagine, for instance, finding a big paper bag full of money in an anonymous urban setting with nobody else in sight. For the first half of my professional life as a lowly paid academic, I found myself occasionally wondering how I would respond if faced with such a dumpster windfall. Would I at least be tempted if it was obviously money lost by criminals and not some poor eccentric's life savings? As we'll see, such purely hypothetical moral dilemmas can be used to scientifically probe the moral functions of our brains, and, as we'll see, among hunter-gatherers in the far north such hypotheticals can even be used to influence how children learn the moral rules of their cultures.

Sometimes, of course, we may simply succumb to life's predictable temptations—in spite of being haunted by an impending sense of shame that combines with fear of punishment. Our consciences not only identify a given alternative as being moral or immoral but also help us decide what to do about it. And in this context it makes evolutionary sense if we can cut some useful corners competitively without taking major social risks. That way, our fitness can be advanced.

So internalization doesn't make people socially perfect. Far from it. But even the opportunistic evolutionary conscience that Alexander identified does serve as a cognitive beacon when we are about to stray and harm our social reputation and also as an emotional inhibitor that often keeps us from straying too far—and perhaps disastrously. Thus, an internalizing conscience has been useful in keeping us out of serious trouble socially, and today in modern society it may also keep us out of jail, where our reproductive success would be considerably diminished. At the same time, a conscience can help us to maintain respect for ourselves, for basically we judge ourselves by the same group moral standards that we use in judging others.

From the previous discussion, it seems obvious that several brain areas have evolved to give us this remarkable moral faculty that might be unique to humans. A sense of right and wrong and a capacity to blush with shame, along with a highly developed sense of empathy, compel us as moral beings to consider how our actions may negatively affect the lives of others—or how we may gain satisfaction in helping them. We also possess the capacity to understand that our groups may punish us for present and past misdeeds, including deeds in the distant past, and a conscience helps us to be aware of our social reputations in general. Yet the conscience also has its Machiavellian functions, for it can guide us to take a flexible approach to being moral that allows us to profit from having a decent reputation and at the same time judiciously cut the occasional not-too-serious corner and profit from doing so.

How should a reproductively useful conscience be designed, then? First, in the Darwinian competition among individuals a conscience shouldn't be too weak because this can lead to personal disaster. Nor should it be too strong, for the internalization of rules shouldn't be too inflexible. We'll be discussing this further, but an efficient evolutionary conscience is one that lets us express ourselves socially in ways that help us to both keep ourselves out of trouble and get ahead in life. For instance, this conscience doesn't make us give the government the benefit of the doubt when we are paying income tax—but it does keep most of us from robbing banks or, unless we are prominent politicians, from committing flagrant and reckless adultery.

CONSCIENCE AND EMPATHY

Ever since Darwin, sympathy and conscience have been talked about in the same breath, but feeling concern for others and listening to an inner voice are far from being identical. For instance, we may stop ourselves from doing harm to another simply because we fear being caught and punished. Of course, we may also feel for our victim and refrain—sometimes even though we have good reason to dislike that victim—because we have internalized a social norm that tells us hurting another human being is wrong.

The interactions of such psychological forces can be complex, and they often breed ambivalence. In *A Fable,* one of William Faulkner's less appreciated novels, soldiers on both sides of the trenches in World War I France mutiny to try to stop the senseless killing.[16] This action is partly a matter of self-preservation, but it is also a matter of conscience and a refusal to dehumanize the enemy. A Christ-like figure is the main actor, and the story involves a combination of motives that include self-preservation, empathy, and morality. Normally, across the front lines in warfare both morality and sympathy tend to be suspended because a soldier is dealing with "outsiders," but Faulkner's well-told story makes clear that under the right circumstances conscience and feeling for others can apply strongly not just to the in-group but to members of the out-group as well.

When people are dealing only with others of their own group, this dimension of the human conscience is prominent.[17] Notions of right and wrong may not *rule* our lives, but they regulate them very significantly, and many of the mores we internalize are shaped by helpful feelings toward others. For instance, in addition to moral rules aimed at keeping us from doing harm to others, there are rules that spur us to give needed assistance to others even if they are not close kin. Clearly, having a sympathetic conscience that includes a sense of shame helps us to fit with the prosocially oriented communities we reside in and to fit in with networks of cooperation that profit ourselves and others. The only problem with having such feelings is that they lead us to aid others even if they won't necessarily pay us back, and this is a major theoretical problem we'll try to resolve in Chapter 7. Again, the answer I'll be favoring is social selection.

LOQUACIOUS HUMANS IN GROUPS

The rules individuals internalize are the cultural product of groups that gossip moralistically on an ongoing basis. That's how moral codes originate, stay in place, and are continuously refined. Surely gossip is a hunter-gatherer universal from way back,[18] and it still manifests itself today in our national media in the form of gossip columns, TV "entertainment shows," and soap operas, while in our workplaces and neighborhoods we continue to discuss privately (and deliciously) the doings of others—just as has been done in small communities of human foragers for dozens and dozens of millennia.

Such "talking" with trusted associates permits people not only to evaluate their peers, but also to intuitively mull over what is useful or disruptive in human social life and to keep in place a moral consensus about how group members shouldn't—or should—act toward one another. In a rather immediate sense, then, it's gossiping that's responsible for the group mores that orient social control and lead people to preach actively in favor of cooperation and generosity. I'll have much more to say on the subject of gossiping in Chapter 9, for every cultural anthropologist has to be adept at this verbal art.

Today human groups come in the form of nations or cities as well as tribes and nomadic bands, but they all have such moral codes. And even though certain types of moral belief can vary considerably (and sometimes dramatically) between cultures, all human groups frown on, make pronouncements against, and punish the following: murder, undue use of authority, cheating that harms group cooperation, major lying, theft, and socially disruptive sexual behavior. These basic rules of conduct appear to be human universals. In any event they are so widespread that we may make an evolutionary assumption that as cultural practices they would have been reasonably well suited to whatever social exigencies were common in Late Pleistocene human living situations, which, as I'll be demonstrating, in many ways were not that dissimilar to exigencies we face today.

It's clear that "biology" and "culture" have been working together to make us adaptively moral. For instance, when we use our cultural acumen to learn moral rules as children, this is based on developmental "windows" that are sequentially hardwired. The same rules about helping others that have been internalized through early child socialization are later reinforced in adults by the prosocial "preaching" we'll be discussing throughout the book, which basically encourages group members to live a socially useful life by being helpful to others. Salient are strictures to be generous both within the family and to group members who are not family. For nomadic modern hunter-gatherers who are strongly egalitarian, we may add being humble in demeanor and avoiding aggressive domination to their particular list of desirables,[19] along with being truthful with other group members, being cooperative and respectful of others, being fair in "business dealings," and being prosocially inclined in general.

If language is used for such encouragement, language also generates criticism that comes in the form of pointedly hostile corrective advice or the mocking of a deviant. And still stronger forms of language-based social control exist, of course, such as group shaming. There's also ostracism or shunning, which conversely removes deviants from normal communication. There's expulsion from the group, a distressing measure arrived at through group consensus. Mobile hunter-gatherers in their small bands do all of these things, all over the world, and very much the same is true of all other humans, whether they're living in larger sedentary "tribes" or in villages or towns or even in enormous urban environments.

In extreme cases where acts of deviance seriously threaten the lives of others or are felt to be truly abhorrent, a death penalty may be inflicted after a hunter-gatherer consensus is reached by privately talking things out. Such dire punishment is still widespread today, and just a few thousand years ago—before movements against such measures arose—it probably was the norm worldwide. Indeed, 15,000 years ago in a Pleistocene world peopled just by mobile foragers, capital punishment surely was universal or quite widespread as a practical but extreme expression of social distancing.

PYGMIES JEER AT AN ARROGANT CHEATER

A vivid instance of hunter-gatherer moral life—and the role of language—can be taken from the Mbuti Pygmies. They live in what was once called, in colonial times, the Congo, a place I have seen only from a distance—and a bit fearfully. For six years I made an annual trip to Gombe National Park in Tanzania to study wild chimpanzees while working in collaboration with Dr. Jane Goodall, and in the early 1980s, from our isolated research quarters on the beach, we could just see the hills of Zaire forty miles away, on the far side of enormous Lake Tanganyika.

Except when storms are raging, this huge inland waterway is easy enough to cross, and on those distant hills some nights the fires burned bright orange for hours, the visible signature of rebel forces resisting the Zairian government and attacking agricultural villages that did not cooperate with them. A few years earlier forty of those same rebels had motored right across the lake to kidnap four of Jane's student fieldworkers, so watching these burning villages definitely brought a sense of malaise. As I heard it, the Stanford University students were eventually ransomed for $500,000 by one student's father, but I knew that if the chips were down, I personally would have no such backer.

The Mbuti Pygmies live in the interior far beyond those hills, in truly dense forests that provide these skillful hunters with enough meat to satisfy their needs because they trade some of it with Bantu farmers for grain the farmers grow.[20] Aside from this unusual economic symbiosis, these foragers live pretty much as other mobile hunter-gatherers do, camping in small egalitarian bands of up to a dozen or so families in one locale, until it is time to move on and exploit another. The Mbuti have no formal religion as we know it, but in their own way they use rituals to worship and placate "The Forest," which, as they see things, provides generously for them. They are a loquacious and intelligent people, and if they are morally aroused, their capacity to express themselves is far from restrained.

Anthropologist Colin Turnbull lived with these people and wrote several books about their way of life. An excellent writer, Turnbull was exceptionally sensitive to the nuances that attend social life in small moral communities. Like myself when I am studying humans rather than chimpanzees, he felt it was important to become reasonably proficient in the native language, and surely, like all cultural anthropologists when a situation of exceptional social interest arose, he took notes as things went along and afterward checked his records independently with a number of indigenous informants to make sure he'd gotten it all right.

The episode I shall be describing here—often using Turnbull's own words and the words of the Pygmies themselves—is one that reflects some of the core moral values in forager life. These involve political egalitarianism and cooperation in procuring and sharing meat, and both practices will play a key role in the coming prehistoric analysis of conscience evolution, just as they play a central role in the social lives of all mobile foragers today.

Here's the shameful story. A rather arrogant mature hunter named Cephu was a member of the Mbuti band in question, even though his extended family seemed to be not as well integrated into the band as the rest of the families, a few of which were closely related by blood but most not. Such an admixture is typical of forager bands, and even though these people are quick to show preference toward their kinsmen, just as Hamilton's kin selection theory predicts, they will in many contexts treat everyone in the band almost like "family."

This is particularly the case when large game—a favorite food because of its exceptional fat content and general nutritious value—is taken down. All over the world, mobile hunter-gatherers use social control guided by moral rules to see to it that when a successful hunter kills a large mammal, his ego is held in check. To this end, he is not only precluded from decisively favoring his family and kinsmen with larger portions of the meat, but usually he is also forbidden by his fellow egalitarians even to preside over and distribute the meat—for fear that he might use this position to gain political or social advantages. Rather, the band sees to it that some neutral person will distribute the meat fairly and equitably, according to the rules.[21] These rules are, of course, moral, and it is virtuous, as well as mandatory, to give over one's kill to the entire group. By the same token, it's dangerously deviant to play the possessive meat bully—and downright shameful to sneakily cheat on the system. This last is precisely what Cephu did.

Foragers most often use projectile weapons to hunt large game in small groups, but the Mbuti sometimes engage in collaborative net hunts that involve the entire band. Each man has a very long net, and up to a dozen nets are positioned so that they form an extremely long, semicircular trap. Some distance away the women and children then start beating the bush and approaching the nets to drive frightened animals like forest antelope into this trap. Each man then spears any prey that become ensnared and keeps the meat for his family.

This variant of Pygmy hunting and sharing does not require a designated, fair-minded meat distributor because the game are medium to small and the nets are so long that everyone is bound to get about the same amount of meat. But that is true only if no one cheats. During the net hunt in question, the egoistic Cephu quietly decided he wasn't getting his due. As fleeing animals randomly rushed into other nets and were speared by their owners, Cephu decided to improve his luck. When he thought no one was looking, in the dense forest he repositioned his net so that it would be well ahead of all the others, and the driven animals would run into his net first. Cephu succeeded very well in the final take, but this cheater had the misfortune of being spotted.[22]

Colin Turnbull had gone along on the hunt, but he was unaware of Cephu's crime, just as Cephu was unaware that his dastardly act had been witnessed. As most of the families were returning to camp, Turnbull noticed a very gloomy mood among them, and he heard both men and women quietly swearing at Cephu, who had yet to arrive. No one would tell Turnbull what had happened, but finally an adult male, Kenge, said to the group, "Cephu is an impotent old fool. No, he isn't, he is an impotent old animal—we have treated him like a man for long enough, now we should treat him like an animal. Animal!"

This statement broke the ice, and some serious gossiping began as the score was carefully added up and a group consensus materialized. The result of Kenge's tirade was that everyone calmed down and began criticizing Cephu a little less heatedly, but on every possible score: the way he always built his camp separately, the way he had even referred to it as a separate camp, the way he mistreated his relatives, his general deceitfulness, the dirtiness of his camp, and even his own personal habits.

Just then, Cephu returned from the hunt. As he stopped at his hut, Kenge shouted over to Cephu that he was an animal! As he strolled over to the main camp, Cephu attempted to tough it out:

> Trying not to walk too quickly, yet afraid to dawdle too deliberately, he made an awkward entrance. For as good an actor as Cephu it was surprising. By the time he got to the *kumamolimo* everyone was doing something to occupy himself—staring into the fire or up at the tree tops, roasting plantains, smoking, or whittling away at arrow shafts. Only Ekianga and Manyalibo looked impatient, but they said nothing. Cephu walked into the group, and still nobody spoke. He went up to where a youth was sitting in a chair. Usually he would have been offered a seat without his having to ask, and now he did not dare ask, and the youth continued to sit there in as nonchalant a manner as he could muster. Cephu went to another chair where Amabosu was sitting. He shook it violently when Amabosu ignored him, at which he was told, "Animals lie on the ground."

Next Cephu was told that he took more help from other band members than he gave back, and Cephu tried to defend himself. It is at that point that Ekianga, another adult male, let on in no uncertain terms that the group knew what had gone on. "Ekianga leaped to his feet and brandished his hairy fist across the fire. He said that he hoped Cephu would fall on his spear and kill himself like the animal he was. Who but an animal would steal meat from others? There were cries of rage from everyone, and Cephu burst into tears."

This action involved very strong shaming, and Turnbull makes it clear that Cephu's deviance was extraordinary: "I had never heard of this happening before, and it was obviously a serious offense. In a small and tightly knit hunting band, survival can be achieved only by the closest co-operation and by an elaborate system of reciprocal obligations which insures that everyone has some share in the day's catch. Some days one gets more than others, but nobody ever goes without. There is, as often as not, a great deal of squabbling over the division of the game, but that is expected, and nobody tries to take what is not his due."

Cephu's next acts were to cover his deviance with a lie and then engage in some egoistic boasting, which, to me, seems almost worthy of the sometimes megalomaniac, fast-talking, recklessly-lying psychopaths described by Hare and Kiehl:

> Cephu tried very weakly to say that he had lost touch with the others and was still waiting when he heard the beating begin. It was only then that he had set up his net, where he was. Knowing that nobody believed him, he added that in any case he felt he deserved a better place in the line of nets. After all, was he not an important man, a chief, in fact, of his own band? Manyalibo tugged at Ekianga to sit down, and sitting down himself he said there was obviously no use prolonging the discussion. Cephu was a big chief, and Mbuti never have chiefs. And Cephu had his own band, of which he was chief, so let him go with it and hunt elsewhere and be a chief elsewhere. Manyalibo ended a very eloquent speech with "Pisa me taba" ("Pass me the tobacco"). Cephu knew he was defeated and humiliated.

Cephu could have continued to protest his innocence and left the band. But he didn't, and Turnbull knew exactly what Cephu had to be thinking.

> Alone, his band of four or five families was too small to make an efficient hunting unit. He apologized profusely, reiterated that he really did not know he had set up his net in front of the others, and said that in any case he would hand over all the meat. This settled the matter, and accompanied by most of the group he returned to his little camp and brusquely ordered his wife to hand over the spoils. She had little chance to refuse, as hands were already reaching into her basket and under the leaves of the roof where she had hidden some liver in anticipation of just such a contingency. Even her cooking pot was emptied. Then each of the other huts was searched and all the meat taken. Cephu's family protested loudly and Cephu tried hard to cry, but this time it was forced and everyone laughed at him. He clutched his stomach and said he would die; die because he was hungry and his brothers had taken away all his food; die because he was not respected.

The playacting was totally ignored, but Turnbull makes it clear that Cephu's apology and concession of the meat set things on a conciliatory path. Within a few hours, Cephu joined the group in its evening singing ritual, and he and his extended family were no longer socially distanced deviants but accepted members of the group again. Reconciliation was, of course, in the interest of all, for it enabled the band to continue to have many hunters and frequent opportunities to eat their beloved meat.

Turnbull, who like myself surely has read famous French sociologist Émile Durkheim on the punitive power of small moral communities, sums up this group-sanctioning episode very nicely:

> Cephu had committed what is probably one of the most heinous crimes in Pygmy eyes, and one that rarely occurs. Yet the case was settled simply and effectively, without any evident legal system being brought into force. It cannot be said that Cephu went unpunished, because for those few hours when nobody would speak to him he must have suffered the equivalent of as many

> days solitary confinement for anyone else. To have been refused a chair by a mere youth, not even one of the great hunters; to have been laughed at by women and children; to have been ignored by men—none of these things would be quickly forgotten. Without any formal process of law Cephu had been firmly put in his place, and it was unlikely he would do the same thing again.

This was a case of corrective social control through shaming and threat of expulsion in which a deviant's behavior was modified so that the group needn't lose a productive member who was misbehaving. The breach of mores was serious, and Manyalibo made one thing clear: if Cephu really thought he was too good to be just another egalitarian band member who followed the group's rules, he was free to take his little handful of relatives and friends, go elsewhere, play the "big chief," and possibly starve. Thus, in trying to defend himself in one transgression, Cephu committed another: he tried to lord it over his egalitarian peers. For both sins he was forgiven, but only after proffering a weepingly submissive apology.

All of these nuances become so obvious because Turnbull offers an unusually detailed description. Cephu's life was never in immediate danger, but the threat of expulsion from the band was a compelling one. People in small moral communities fear the wrath of the group for good reason. If a crime is an ultimate one, and if words and social pressure are not sufficient to rehabilitate the transgressor, sanctioning, both physical and verbal, can become far more decisive. Although some hunter-gatherers don't do this, the Mbuti beat sneak thieves when they are caught. And any small human group has the potential to use capital punishment if a deviant poses a sufficiently ultimate danger. But even just being expelled from a band can bring serious risks, and it was in the face of such a threat that Cephu came down off of his defensive high horse, grudgingly but apologetically all but admitted his shameful act, and, having submitted, became an accepted group member again.

THE EVER-PRESENT THREAT OF RIDICULE

The impact of actively shaming deviants deserves further discussion. Once ridicule has been used to shame someone, say for behaving arrogantly, others with a similar penchant are likely to stay in line almost automatically—just to avoid similar humiliation. More than fear is at work here, for the potential deviants have internalized egalitarian group mores that condemn self-aggrandizement, have personally experienced shame feelings while growing up, and fear being further ridiculed or shamed as adults. They also have language, so the learning of moral lessons can be vicarious. Any Pygmy who later heard the story of Cephu would think twice before shifting his net ahead of the others.

The effectiveness of even mild forms of ridicule is perhaps best seen in anthropologist Richard Lee's vivid (and oft-quoted) descriptions of how the !Kung Bushmen keep alpha-male tendencies in check. (The "!" that precedes "Kung" is a clicking sound that is part of their language.) Unlike the Mbuti, these Kalahari Bushmen are economically independent mobile foragers of the same general type as people who were evolving our genes for us 45,000 years ago, and they, too, are verbally adept. When a !Kung hunter comes back from a hunting expedition, others in the camp are eager to eat their favorite food, meat, and expectantly they'll ask him what he has killed. Knowing that he'll be subjected to ridicule if he shows the slightest tendency to boast and set himself up as being a superior hunter, he'll all but poetically deprecate the size and quality of his prey. An articulate Bushman named Gaugo tells Lee, "Say that a man has been hunting. He must not come home and announce like a braggart, 'I have killed a big one in the bush!' He must first sit down in silence until someone else comes up to his fire and asks, 'What did you see today?' He replies quietly, 'Ah, I'm no good for hunting. I saw nothing at all . . . maybe just a tiny one.' Then I smile to myself because I know he has killed something big."[23]

Or as a renowned healer named Tomazho says, "When a young man kills much meat, he comes to think of himself as a chief or a big man, and he thinks of the rest of us as his servants or inferiors. We can't accept this. We refuse one who boasts, for someday his pride will make him kill somebody. So we always speak of his meat as worthless. In this way we cool his heart and make him gentle."[24]

Thus, even though the successful hunter's chest may be quietly swelling with pride, he'll shape his words very humbly, and his egalitarian peers, all too ready to put him down with ridicule, will approve his self-effacement and respect him both as a hunter and as a person of humility.

Cutting proud hunters down to size verbally isn't the end of it, for usually Bushmen don't even get to distribute the meat they've hunted. Once the carcass is hauled into camp, by custom someone else will probably preside over the meat and share it out to the main kin groups in the band—who'll then share it further with their close kin and other associates. The effect is to remove the hunter from the meat he has killed as a possible ticket to power, and the Bushmen understand this situation all too well.

WAS CEPHU A PSYCHOPATH?

This beautifully detailed pair of ethnographic descriptions from sub-Saharan Africa shows how groups morally manipulate individuals for practical purposes that serve everyone but the would-be deviant and his family. With the Pygmies we saw how morally based indignation can spread contagiously through a group so that almost everyone becomes emotionally exercised—even though a few persons may take the lead. It also shows that the close relatives of the culprit may stand aside and remain neutral, for Cephu's extended family didn't verbally attack Cephu or join in the shaming.

But even though they didn't enter into the active criticism, ridicule, shaming, and threat of expulsion directed at Cephu, they were not trying to defend him either. Had they done so actively, we simply would have had a case of conflict within the group, with both sides trying to use morality to justify themselves. Instead, we had an instance of moral sanctioning in the name of the group and its vital social functions, with a familial faction choosing to stand aside rather than backing their leader, who was so clearly in the wrong.

Cephu did try to defend himself, but we must assume that he understood all too well how repugnant meat-cheating was to his bandmates and that to some significant degree he'd internalized group values that condemned such behavior—unless he was a full-blown psychopath. His argument that he was a "big man" who need not follow the rules was extremely repugnant to his Pygmy colleagues, just as it would be in any mobile forager band, with its emphasis on the essential equality of all the adult hunters. And it smacked of the grandiosity that we see in American psychopaths who are incarcerated. But all that line of reasoning got him was a threat—that if this was the case, then he could take his extended family with him and split from the band.

It's tempting to suggest that Cephu might have been *something* of a psychopath, and in studies of modern psychopaths this affliction has been found to be a matter of degree. However, this would be very difficult to determine because we lack studies for Mbuti Pygmies like those of Hare and Kiehl. Furthermore, in spite of the sometimes arrogant and facile attempts to defend and justify his behavior, Cephu did *appear* to be engaged with normal moral emotions even though their expression was combined with playacting.

If manifestations of psychopathy hold across cultures, which seems very likely, Cephu may have had at least a touch of this innately based moral ailment. This is sheer speculation, however, and privately his remorseful feelings afterward could have been deep or shallow or even nonexistent. We'll never know, unless Dr. Kent Kiehl decides to move his mobile MRI wagon to the forests of central Africa.

MOBILE FORAGERS AND THEIR SOCIAL CONTROL

Let's consider more broadly the nature of hunter-gatherer social control. To do so, we must move from this pair of African societies to the many scores of anthropologically studied mobile band cultures, which, like the Bushmen but not like the grain-bartering Pygmies, are directly comparable to the independent mobile bands that lived under Late Pleistocene conditions. Before the Holocene Epoch phased in about 10,000 years ago, this prehistoric world was populated mainly or possibly exclusively by politically egalitarian hunter-gatherers, and because an individual lived in a small band of about twenty to thirty or perhaps forty people, she or he certainly didn't want to get on the wrong side of the group as Cephu did.

Group moral indignation can take a number of forms, most of them quite uniform today among these foragers from one continent to the next. Their reactions range from moderate rebukes and sharp criticism to ostracism, ridicule, shaming, and outright banishment; and at the end of the line is the fearsome specter of capital punishment. Foragers—who morally appreciate the sanctity of human life within the group and do so strongly—use this measure rarely but decisively, as a desperate last resort. Presently, in Chapter 4 we'll be seeing which crimes bring about this dire type of community reaction.

I've already drawn up a short list of proscribed behaviors that seem to be universally condemned and punished. However, the actual strength of these group reactions can vary from culture to culture. For instance, even though close degrees of incest are universally disapproved, in some band-level cultures such behavior is not punished very severely, perhaps by a scolding or by ostracism for a time, whereas in others it brings a death sentence because it is felt to be so monstrously abhorrent or so threatening to group social life or to the lives of other group members.[25] When it comes to serious meat cheaters like Cephu, however, they are always likely to be treated roughly because their behavior is seen as threatening everyone else's welfare, and in some groups they may be killed. Likewise, if a man becomes overbearing in dealing with his peers or, where shamanism prevails, if a person selfishly and maliciously misuses supernatural power, that deviant may be actively rebuked or the rest of the band may simply slip away in the night. But such people may also be killed if there's no other way to escape their threatening domination.

In fact, if such self-aggrandizers try persistently to intimidate or tyrannize their peers, or if they actually succeed in doing so, they are quite likely to be killed. Indeed, with these egalitarians, to seriously disrespect other hunters in the group and to trample on their precious rights as equals creates really serious anger and disapproval, leading to true moral outrage. That said, however, foragers take little joy in killing a group member, and usually they try to reform deviants rather than eliminating them through banishment or execution.[26] This is partly because they feel for them as fellow human beings, and partly because they're practical people who understand the need to have as many hunters as possible in the band—even if, like Cephu, they have irritating qualities and are occasionally prone to deviance.

As we'll see in Chapter 4, however, when it comes to the really *serious* political dominators I just mentioned, and also a few other seriously deviant types, a firm and sometimes ultimate line is drawn in the sand: he who crosses this line must be prepared to sacrifice his genetic future.

Of Altruism and Free Riders 3

THERE ARE MANY RULES THAT ARE GOLDEN

Religious versions of the Golden Rule are being promulgated and preached widely today,[1] and some such stricture is likely to surface in any cooperative human group. These come in the form of secular exhortations that remind people to do good and avoid doing harm because "what goes around, comes around." For instance, consider the following dicta and their apparent universality:

> *Do unto others as you would have others do unto you.*
> —Classical Christian statement of the Golden Rule

> *Hurt no one so that no one may hurt you.*
> —Muhammad, The Farewell Sermon

> *Never impose on others what you would not choose for yourself.*
> —Confucius, Analects XV.24

The active promotion of reciprocity in everyday social life appears in bands and tribes and also in hierarchical chiefdoms; indeed, every human society has this. These sayings promote generosity, and the underlying intentions are always the same. Obviously, they are prosocial: your parents, friends, and neighbors want you to behave generously and with an eye to future reciprocation. It's for good reason that cultures so predictably develop such "preaching," for the underlying assumption is that in society at large generous reciprocation needs badly to be fostered—and that in being human, people will need some serious prodding in this direction.[2] The desired result will be more cooperation—and less conflict—because generosity tends to beget more generosity.[3] The name of the game is indirect reciprocity.

GOLDEN RULES AND INDIRECT RECIPROCITY

We'll meet with some exact figures for hunter-gatherers in Chapter 7, but this prosocial preference appears to have been very widespread and possibly universal among the recent human foragers who were evolving our genes for us.[4] As we'll see in the next chapter, in practice these people all share their big-meat carcasses in a reasonably equitable fashion and manage to do so even when some hunters are far more productive than others. This goes on even though families are constantly changing bands, which means that anything like exact reciprocation is out of the question.

If bad luck arrives, band members will, within limits, help others in the band according to need—but even though any person of good or acceptable moral standing is automatically eligible for these indirect reciprocity benefits, a few very large beans will, in fact, be counted. In general, individuals whose track records are unusually generous may receive more help than those who have been stingy. And as for people who have a long record of being outlandishly and immorally selfish or lazy, help from unrelated group members may simply be denied.[5] When personal trouble strikes, and with only their close kin likely to support them,[6] such individuals may wish they had followed the Golden Rule.

Coming up with admonitory rules calling for generosity within the group is a cultural way of massaging any contingent system of indirect reciprocity.[7] Having an efficient conscience makes possible the internalization of such rules, and even though this certainly doesn't guarantee the conduct, it serves as a constant brake on selfishness and as a spur to being generous. In a hunting band these rules are important to everyone's physical welfare, for generosity in everyday life is centered on sharing a highly nutritious food. And this routinized meat-sharing goes far beyond satisfying feelings of fairness and a love of meat, for as we'll see, it enables an entire, interdependent hunting band to maintain vigor and health. Since all large game is shared, this benefits every worthy individual in the group and also the group as a whole.

For foraging nomads who seldom invest in storing food, meat-sharing amounts to a system of insurance,[8] and this ancient type of risk reduction was carried forward into the Holocene Epoch and the era of domestication, even as human social forms were changing because populations were becoming denser. For instance, in a sizable hierarchical nonliterate chiefdom in which agricultural food storage is practiced, a major portion of a household's annual produce is given to the privileged chief. The chief will set most of it aside and then, over time, will give it back to those in need.[9] This could be called a centralized version of contingent indirect reciprocity,[10] and it can lead to still more centralization. Starting with the earliest civilizations, we find formal systems of taxation that are no longer voluntary,[11] but even with a coercive centralized government in control, the prosocial pronouncements continue in force—as a means of reinforcing generous behaviors that will help any overall system of governance by fostering cooperation and reducing conflict.[12]

No matter what size the society, people everywhere seem to realize that by reinforcing and amplifying individuals' tendencies to extrafamilial generosity,[13] they can improve the overall efficiency of cooperation from which everyone profits. At the same time, they understand that failure to reciprocate can cause conflicts that seriously disturb group social or economic life. In human minds everywhere, prosocial generosity is good, inappropriate selfishness is bad, and conflict is to be avoided.

Even among hunter-gatherers such moralizing pronouncements go beyond individuals merely nagging other individuals in a very immediate, self-interested way; in a sense these people may be seen as intuitive applied sociologists who are purposefully trying to shape their society in ways that will help themselves because everyone's life is helped by better cooperation. Similar social creativity continues today with modern safety nets as we work to create systems of insurance against subsistence shortfalls, illness, or injury, the huge scale of which would have been unimaginable a mere 10,000 years ago.

The general idea behind all of these calls for generosity is to get people to contribute more willingly and more predictably to the shared social and subsistence life of the group as a whole. Of course, modern insurance systems are so formalized and bureaucratized that this voluntary element gets lost in the shuffle, but today we also have major "do-gooding" industries that are, in fact, based on altruistic good intentions. These appear in the form of secular and religious nongovernmental organizations, in which stimulating people through golden-rule-type pronouncements is needed because the contributions are totally voluntary and often anonymous. Goodwill Industries is one American example out of many, in which a small measure of donor effort will result in major benefits for others. And as with the Red Cross, the name is designed to promote such effort.

In short, encouraging people to give generously—and then to receive contingently if in need—has both a venerable past and a solid future, and it serves as a culturally invented antidote to the predictable effects of human "selfishness." This insight originally came from my mentor, the late Donald T. Campbell, who was fascinated by the social tensions produced by a human genetic nature that is, at the same time, significantly generous and immensely selfish.[14] Only a species with a powerful, socially oriented brain and language could come up with such prosocial "propaganda," for promulgating these idealized rules involves an active "functionalist" view of society, taken as a working system that can be deliberately enhanced to increase the welfare and security of all.[15]

Such manipulative preaching can work quite well with a species that is genetically evolved to internalize rules, and in this context ethologist Irenäus Eibl-Eibesfeldt has spoken of the general "indoctrinability" of humans.[16] Golden rule promulgation is a prime instance of group members realizing that others can be swayed by words—for the public good—and at the base of this process lies the evolutionary conscience as a judgmental, trainable cultural sponge that is shaped early in childhood. When we consider the Kalahari Bushmen and the Inuit of Alaska, we'll see that parents place generosity on the positive side of the ledger when children are quite young. Golden-rule exhortations do the same thing for adults, as people try to shape their social life by campaigning against undue selfishness.[17]

As we'll see in Chapter 7, the social encouragement of generous behavior promotes generous reciprocation not only among a band's different families but also within the family. Among unrelated families in a band, such "tweaking" is needed because human altruism is weak compared to nepotism. Within the family, similar "tweaking" is needed because generally speaking human nepotism seems to be weak compared to the basic egoism that is natural selection's gift to us. Band members call for generosity in both contexts because both family and nonfamily squabbles interfere with cooperation and disturb everyone else in the band, and the threat of a major conflict's splitting the group is always on the minds of these foragers.

WHY ARE WE SO EGOCENTRIC?

Today, the scientific selection basis for explaining the strongly egocentric side of human nature seems just as solid as Darwin supposed it to be well over a century ago, when he created his competitively based theory of natural selection. Initially, this remarkable theory was totally "egocentric" in its orientation, but, as we've seen, as Darwin pondered certain generous tendencies that failed to fit with this powerful new theory, he came to realize intuitively that kin selection makes it easier for us to be generous toward our closer blood relatives. He also realized that the broader, extrafamilial type of generosity, which these golden rules are so regularly designed to reinforce, required a different explanation.

Darwin's gene-less but basically heredity-based evolutionary logic had drawn him toward the group selection type of theory that we've already touched upon at several points,[18] but Darwin had no way of understanding group selection's mathematically predictable mechanical weakness compared to selection taking place within groups. Nor did he have any way of recognizing the enormity of the free-rider problem,[19] which today in most scholars' minds stands as a serious general obstacle to the evolution of wider generosity in any mammalian species—aside from naked mole rats, who essentially live in great big isolated clans.[20] Yet in his own way Darwin did astutely identify very much the same paradox of extrafamilial generosity in 1871 that Edward O. Wilson redefined for us so influentially over a century later.[21]

Today we know what a gene is and we know how to model mathematically what is likely to be taking place in gene pools. But with all this knowledge, and after almost four decades of diligent and inventive research, there still seems to be no single satisfactory answer for the puzzle of humans being able to behave as generously and cooperatively as we often manage to do. A basic question is, how did natural selection manage to work its way around the powerful degrees of genetic egoism that are built into our nature?[22]

In this chapter we'll be considering the major theories that have been put forward over the years as evolutionary scholars have tried to account for extrafamilial generosity in humans. Highlighted will be a special, social selection venue for explanation, a theoretical path that builds on established paradigms but also involves some new elements that, I believe, will make generous human responses to the needs of others much easier to explain. An important new element will be the systematic, punitive suppression of free-riding behavior, which includes not only curbing the predatory actions of cheaters, but also suppressing the selfish exploits of a different type of free rider, an intimidator who is very different, indeed, from those normally considered. In this context, thoroughgoing suppression of the powerful has become prominent in egalitarian humans and among them alone, as will be seen in a later chapter. I believe this to be a main reason for our being as altruistic as we are.

EMPATHY AND GENEROSITY

When evolutionary scholars speak of altruism, even in the strict sense of extrafamilial generosity this can still have a variety of meanings. One is purely a matter of genetics: you give up some of your own fitness to increase the fitness of a nonrelative. That's basic. However, when emotions and motives enter into the picture, things can become more complicated. For instance, you may give to another because you expect an immediate or eventual return; you may give because you fear gossip and public opinion; you may give because seriously unacceptable nongiving can bring active punishment from your peers; you may give as a social conformist just because that's what people do in your culture, and this makes for an easy conscience.[23] And, of course, you may also give in a heartfelt way because you identify with another's need or distress and it just feels right to be helpful.[24]

Obviously, this last kind of giving is based heavily on sympathy, an emotion that leads us to care mainly about those we're socially and emotionally bonded with. Although "empathy," as technically defined by psychologists,[25] seldom figures in more formal anthropological analyses of hunter-gatherer cooperation, fortunately there's one study that does indirectly take this aspect of generosity into account, and we'll be meeting with it later in Chapter 11.

For now, let me say that in today's small hunting bands, people form substantial social bonds with most group members, be they blood relatives or otherwise, and there can be little doubt that these positive associations invoke the sympathetic kind of feelings that Darwin emphasized,[26] which make it possible for one person to emotionally identify with another's needs and provide help accordingly. Help to close kin is readily compensated genetically by kin selection, but among nonrelatives major donations will involve substantial net fitness costs that somehow need to be compensated—hence, the genetic puzzle we keep speaking of.

HOW "LOOSE" IS THE PROCESS OF NATURAL SELECTION?

Apparently, natural selection hasn't managed to set up very effective barriers to cut off such generous helpfulness whenever it is being directed at nonkin. Were egoism and nepotism the only forces driving natural selection for humans, and were the processes involved totally efficient and totally determined by biology, we would expect something quite different. Indeed, first we would expect ourselves to have evolved some foolproof means of identifying kin so as to avoid donations to nonrelatives—and also of making sure we helped our kinsmen only in proportion to the degree of mutual relatedness. And second, we should have evolved to never give costly, uncertainly reciprocated help to nonkin, for such "genetic self-sacrifice"[27] defies the very notion of the efficient natural selection process that evolutionary biologists assume to be operating when they create mathematical models of gene selection.

George Williams, himself a biological mathematical modeler par excellence, describes forcefully how natural selection's lack of total efficiency could occur even with respect to something as basic as reproductive behavior, saying that "reproductive functions, perhaps to a greater extent than any other adaptations, are characterized by a considerable degree of looseness in timing and execution."[28]

In this context, he notes that homosexual behavior is widespread among animals. If such looseness applies also to benevolent helping of nonkin, Williams predicts that the help to "unrelated animals should never be more intense, and should usually be less intense, than the same behavior toward offspring."[29] Thus, in effect, the very substantial benefits of kin selection could be genetically subsidizing the occasional acts of wider generosity that go astray. Williams does not propose that this type of "misplaced reproductive function" could be the sole explanation for altruistic helping behavior, but he does point out "that when an animal actively assists an unrelated individual, it uses only those behavior patterns that are seen in a family setting."[30]

Although intriguing, misplaced reproductive effort is not mentioned very often as a possible solution for the mystery of altruism in human or other species. Although thousands of evolutionary scholars have been struggling to resolve this paradox for these past four decades,[31] they've done so mainly through theories involving very efficient mechanisms that directly *compensate* altruists for the losses inherent in being altruistic.

The effort to explain extrafamilial generosity as an important component of human cooperation continues in many directions, but I shall begin with this "looseness model" that I have just introduced.

1. "MISTAKING" NONKIN FOR KIN

Kin selection is a powerful agency that can sustain generosity to blood relatives in accordance with the strength of the blood tie, and this model readily accounts for a fair amount of the generosity seen in a hunter-gatherer band that usually consists of about 25 percent close kin.[32] What the model doesn't account for, however, is all the generosity shown to nonkinsmen—unless, somehow, some of this generosity is simply "spilling over" owing to natural selection's inefficiency as Williams suggests. This might be called a *slippage model,* meaning in the case of altruism that overall the individual advantages of nepotistic generosity are so strong that *moderate* amounts of costly extrafamilial generosity could be "piggybacking" on them genetically,[33] with no *major* harm done because the inclusive-fitness advantages of nepotism are so strong and the costs of being altruistic are weaker.

For humans, one immediate agency that could facilitate such slippage is our cultural inventiveness in assigning people to social categories. By custom, people in bands sometimes use primary terms reserved for close kin, such as "mother," "uncle," "sister," or "brother," to refer to distant kin or to nonkin they are closely bonded with. If we assume that the use of such terms summons up sympathetic feelings, generosity to unrelated others could come into play because in effect band members are "tricking" themselves in ways that redirect their sympathetic generosity from kin to nonkin. In fact, you may be far more closely bonded with a generous nonrelative with whom you spend a lot of time collaborating in subsistence activities than you are with your own selfish cousin, whose niggardly habits tend to rub you the wrong way. In that case, you're more likely to give assistance to the nonrelative in time of need even though, at the level of genes, you're not being nepotistically compensated through inclusive fitness.

The genes that make this "imprecision" possible might be considered multipurpose or, technically, "pleiotropic."[34] This explains how some moderately maladaptive assistance to nonkin can "piggyback" on the highly adaptive assistance to kin. However, as Williams suggests, over evolutionary time this slippage-based beneficence can continue only if these two behaviors in combination bring a net relative-fitness benefit to the generous individuals involved.

2. CULTURAL DOCILITY AND GENEROSITY

Economist Herbert Simon's docility model provides quite a different kind of piggybacking possibility.[35] This model requires no sympathetic feelings, at all. In Simon's conception, there are enormous advantages stemming from more "docile" individuals being better set up to automatically copy useful behaviors from other culture members without having to engage in costly trial-and-error learning. For example, out on the Kalahari Bushman parents tell their children exactly where poisonous snakes are likely to be encountered. This lesson is best learned vicariously, and I was grateful to learn it that way when I traveled to central Africa to study conflict management among wild chimpanzees. (My personal and dreaded serpentine "favorite," by the way, was the hyperaggressive black mamba, followed by spitting cobras.) When I arrived at Gombe, one of my first questions was about snake hazards, and the long list Jane Goodall told me about also included green mambas, boomslangs, night adders, water cobras, nonpoisonous but seriously sizable pythons, and tiny vine snakes that can barely bite but will kill a person with nerve poison if they do. There were just enough deaths on record to make it obvious that internalizing this information in advance was far better than relying on trial and error, and my cultural docility, combined with a truly major fear of snakes, helped me to stay safe.

Consider now that among all the personally useful cultural patterns that group members pass on to the next generation, there will be pointed, golden-rule-type messages that call on individuals to be generous even to nonfamily. When a person automatically acts on such messages, the moderate altruistic costs will subtract from the very substantial overall individual benefits that come from being generally so spongelike with respect to cultural learning. But as an innately absorptive learner, the person will still be making a net gain.

Of course, "nonconformists" who inherit faulty sponges can easily resist such messages to be altruistic, but they will also be losing out on the general advantages of conformity. For example, they're more likely to die of snakebite. On the other hand, someone who is able to internalize all the useful rules of living but resist the cultural messages that promote altruism could be a free rider on this system.

3. BEST GROUPS WINNING

As we've seen, group selection provides a very different kind of explanation, and it need not involve the altruistic individuals' somehow being compensated. Were group selection strong enough, it could straightforwardly support individually costly, empathy-based, group-useful cooperation among nonrelatives in the same band[36]—especially if all the band members were staying put for life.[37]

Earlier opinions in evolutionary biology held that group selection could work only rather feebly,[38] but there was no disagreement that its effect would have been to contribute to the evolution of sympathetic tendencies that help to motivate genetically costly generosity toward anyone in the group, be it nonrelatives or relatives. Just as Darwin said, this can lead to more surviving offspring in high-altruism groups than in their competitors, which exhibit less sympathy, less generosity, and, at the bottom line, less cooperativeness.

One major argument of anti–group selectionists has been that not only is group selection inherently weak, but the models also are highly vulnerable to free riding.[39] With respect to inherent weakness claims, technical simulation work by economist Sam Bowles[40] demonstrates that for humans group selection could have been a major force prehistorically, because of major genetic differences between groups. With respect to free riders, from the standpoint of modeling the evolution of altruism they are a nasty piece of work, as it were. In past mathematical modeling they are basically "cheaters" who are designed to take advantage of gullible, vulnerable altruists by deceptively taking without giving.[41] Thus, these freeloaders can cash in on the benefits of cooperation without paying any of the costs, which means that as individuals they will easily outcompete the altruists, whose genes thereby lose out and—in theory—all but go away. If this free-rider problem could be eliminated or seriously ameliorated, group selection's power to support extrafamilial generosity would be increased. The arguments in this book will point in precisely that direction, not only for group selection but also for several other models I'll be discussing here.

4. RECIPROCATED "ALTRUISM" AS A THORNY QUESTION

Biologist Robert Trivers's reciprocal altruism model is quite beautiful, in its long-term, "tit-for-tat" symmetry. It's appealing because in theory this model might account for a great deal of the extrafamilial generosity we see.[42] However, this would be so only as long as the unrelated pairs involved were cooperating consistently over the long run, and only if certain other conditions were being met, namely, the exchanges would have to be reasonably equilibrated, which means no major cheating.[43]

Given my definitions, there's a semantic problem with calling such balanced reciprocation "altruism," in that neither party is paying any special costs; indeed, they are both coming out ahead in comparison to individuals who do not pair up and thereby forfeit the advantages of cooperation. However, the real problem is that indirect reciprocity—which is what cooperative foragers actually practice as they help others in need or share their large game—is far from being either dyadic or anywhere near exact in the relative long-term contributions of different individuals or households.[44] In fact, the amounts of communal meat that individual hunters provide to their bands over a lifetime can vary quite substantially.[45]

Where might this elegant and seductive, game-theory-based model of Trivers's apply, then? In humans, probably the closest thing to a real-life relationship in which major contributions of two unrelated partners can equalize out over time and bring major mutual benefits would be a stable, lifelong marital-procreative arrangement. In the absence of sexual cheating, enormous and quite equal reproductive benefits automatically accrue to both parties in the form of offspring. The main type of free-riding deception that would throw this kind of reciprocity seriously out of balance would be cheating by females, because this can result in the male partner's investing heavily in parenting his unrelated genetic competitor's child. Marriage also involves economic reciprocation, and with most of the prehistoric types of foragers we'll be meeting with in the next chapter, husbands and wives make rather different but overlapping types of contributions to the family economy; this means that the costs and benefits would be cognitively difficult to balance even if the two unrelated partners were trying to count every last bean—which in fact they aren't.

It's curious that so little has been written about this very special, sexualized, two-person version of reciprocated extrafamilial generosity, for whether a perfect give-and-take equilibration exists or not, marriage partnerships can bring very large reproductive advantages compared to not pairing up. As a preferred way of doing things, procreative pair bonding appears to have been universal among humans at least since we became culturally modern, and given the likely payoffs, its contributions to our altruistic potential might have been significant.

Trivers's famous model has been employed optimistically by many scholars to explain human cooperation in a variety of other contexts, where the fit with everyday behavior is, I believe, far less compelling. However, the use of his model to explain the affectively warm, generous, sexualized pair bonding of marital partners seems quite promising, and I think this merits further exploration. Obviously, free-riding issues, such as laziness or sexual cheating, do pose problems, but it's worth noting that in general small foraging groups universally frown on major laziness or adultery and sometimes, at least, punish them harshly. In addition, divorce appears to be a forager universal, and it offers some protection against cheating.

5. IMMEDIATE, WELL-EQUILIBRATED MUTUAL COLLABORATION

Short-term mutualized collaboration resulting in balanced benefits involves two partners of the same species engaging simultaneously in one-shot cooperation in contexts so immediate that cheating can simply be set aside as a free-rider issue.[46] Cooperation based on mutualism does occur in real life, as when two African foragers are smart enough to quickly gang up against and bluff away a big-cat predator that could easily have taken either person had each acted solo. As a nice example, the Hadza of Tanzania do their night hunting around water holes in pairs because lions will pick off single hunters.[47]

Today, one-shot mutualistic approaches to explaining cooperation in various species have largely replaced Trivers's very demanding long-term dyadic model, but in my opinion their potential for explaining ongoing extrafamilial generosity in human foragers seems limited. This is because the nutritionally important indirect reciprocity actually practiced by foragers is so far from being immediate or restricted to dyads or balanced; indeed, forager contingent meat-sharing and security-net systems last over lifetimes and involve bands of several dozen people—with families constantly changing bands.

6. SOCIAL SELECTION AND GOSSIP-DRIVEN PREFERENCES

It was in this socially flexible context that biologist Richard D. Alexander influentially coined the term "indirect reciprocity,"[48] and he emphasized superior outcomes in mate choice as a major mechanism that could be supporting the substantial extrafamilial generosity involved in helping those in need according to one's resources. The idea is that being generous makes you look competitively attractive as a partner in a situation of cooperation, be it marital or otherwise. Thus, when others choose you preferentially over someone else who does not display such generosity, this is good for your relative fitness, and the costs of being generous can be more than compensated by benefits coming from being chosen more readily as a partner in cooperation.[49] The idea is that people who are able to partner up, and partner up well, will outreproduce those who fail to do so.

Alexander did see cheating as a serious potential problem in that people could manipulatively showcase or even dissemble generosity. This means that cheating free riders would present problems here, just as they did with group selection and reciprocal altruism. But cheating aside, if it were strong enough, selection by reputation could go far in explaining the often highly contingent type of generosity that is inherent in social systems based on indirect reciprocity—and this is exactly what hunter-gatherers actually practice when it comes to sharing large game or when it's time to help those who are sick, injured, snakebit, or otherwise seriously unlucky and in need of help.

In 1987, Alexander's big-picture "modeling" was anthropologically down to earth, meaning that it had the advantage of being keyed to actual behaviors in the same type of people who most recently evolved our genes for us. His favorite extant group was obviously the exceptionally well-studied Kalahari Bushmen and the !Kung in particular. In more recent years the selection-by-reputation model that Alexander formulated has been explored in the laboratory, using mainly student-based, game-theory experiments, while indirect reciprocity also has been investigated out in the field anthropologically, sometimes by examining meat-sharing systems that accomplish variance reduction, sometimes by focusing on "costly signaling,"[50] and sometimes by studying social behavior that provides social safety nets.[51]

People's reputations are determined by what others see them doing, but even more by their being talked about. Language permits individuals in small groups to exchange such firsthand and secondhand information, the result being a thorough and very useful general knowledge of people's reputations. Not only are good reputations known, but also bad ones,[52] and both are taken into consideration when social choices are being made. For example, a person who is unusually generous would be given some preference as a subsistence partner or marriage partner, whereas an unusually selfish individual who is prone to bullying, cheating, or theft might be carefully avoided by those in a better position to choose.

Add all of this up, and selection by reputation appears to have been a powerful agency in shaping certain behavioral aspects of the human gene pool. As a mechanism similar to Darwinian sexual selection, it favored costly traits involving self-sacrificial generosity but also others that were not costly. For instance, being personally dependable may not be costly, but it is an attractive trait. Also attractive is being hardworking, which in fact would be highly useful to individual fitness regardless of whether an individual was in a position of being chosen or not. Again, the theoretical joker in this selection-by-reputation pack would be free riders, who are able to fake the desired qualities and thereby make themselves as attractive as the real good guys.

SUPPRESSING FREE RIDERS

These six hypotheses for the support of extrafamilial generosity have been widely debated, but the free-rider suppression we're about to discuss brings a new way of approaching the altruism paradox. For humans, the active, punitive social suppression of free riders does not directly select for altruism; rather, it disfavors these classically deceptive born enemies of altruists by either completely suppressing their behavior at the level of phenotype or by placing them at a net genetic disadvantage. These effects open the way for mechanisms such as selection by reputation or reciprocal altruism or group selection to support altruism more effectively.

Over a decade ago, I touched upon this free-rider suppression effect in discussing Late Pleistocene possibilities for group selection to operate,[53] but the idea deserves a much more extensive treatment. As modeled, these notorious classical freeloaders are "designed" to be efficient and insuperable predators in that they are innately prone to take advantage of their more generous peers by actively deceiving them or by failing to reciprocate by standing aside.[54] It's partly for this reason that the great majority of evolutionary scholars have been unwilling until quite recently to give any serious consideration to group selection theory.[55]

In evolutionary theorizing as in everyday life, if you happen to be investing your money with the likes of a Wall Street Ponzi-schemer like Bernie Madoff, the importance of free riders to your welfare, and to your overall fitness, can be enormous. With respect to Alexander's selection-by-reputation hypothesis, we've just seen that it, too, is vulnerable to free riding if poseurs can simulate usefully attractive qualities of generosity.[56] Likewise, in marital unions deceptive adulterous free riding (especially by females) can be a major obstacle to reciprocal altruism's being balanced well enough to work very strongly. In any context of cooperation, a deceptive partner or even just a very lazy partner is problematic not only in an immediate sense but also with respect to your fitness. Indeed, individuals who are "designed" to take more than they give pose a major problem for modeling altruistic genes and their chances of reaching fixation in human gene pools.

I believe that this question of selfish free riders requires further and critical thought and, furthermore, that selfish *intimidators* are a seriously neglected type of free rider. There's been what amounts to a single-minded focus on cheating, which has dominated free-rider theorizing ever since Williams and then Trivers made these antisocial defectors famous.[57] In fact, in human evolution I believe the more potent free riders have been alpha-type bullies, who simply take what they want.[58] A great deal of this book will be about bullies and what small groups of people do about them.

Basically the free riders identified by George Williams are selfish opportunists, tricksters who are evolved to exploit a generous individual's vulnerability to their own genetic advantage. A bully can fulfill this role as well as or better than a cheater. Bullies obviously have no need for deception, for bald use of force (or the threat thereof) is their métier, and any decisively hierarchical species is subject to significant free riding of this type. This means that generally the selfish alpha-male types (and wherever they appear selfish alpha-type females as well) can be very big winners.[59]

High rank, if it can be freely expressed through selfishly aggressive dominance, pays very nicely in fitness, and those who lose out in this competition are not necessarily less physically powerful. They also are likely to be relatively generous, or tentative in asserting themselves, which makes it likely that a significant proportion of the victims of high-ranking aggressors will be altruists. As we'll be seeing in Chapter 5, among our distant ape ancestors this bullying type of free riding was strongly in effect because basically this ancestor lived in social dominance hierarchies, not in egalitarian societies.[60] With humans, however, things have been quite different.[61]

SOCIALLY NEUTRALIZING THE WOULD-BE BULLY

It's here that my work on the evolution of hunter-gatherer egalitarianism comes in, namely, the emphasis on the active and potentially quite violent policing of alpha-male social predators by their own band-level communities. I'm speaking of large, well-unified coalitions of subordinates and their aggressive and effective control of selfish bullies, whose predatory free rides at the expense of less powerful or less selfish others could otherwise be easily taken by force. In the next chapter, we'll see that 45,000 years ago very likely almost all the humans on this planet were practicing such egalitarianism.

Just as cheater detection and cheater avoidance by individuals can reduce the advantages of free-riding con artists,[62] I've shown that humans' collective antihierarchical sanctioning can behaviorally neutralize, and sometimes reproductively penalize, these otherwise unstoppable bullies.[63] When such penalties come into play, this also intimidates other would-be dominators, and overall the winners are those who are less disposed to use power selfishly, including altruists whose competitive tendencies are tempered by generosity.

Let me preview something else, now, about these culturally modern predecessors of ours. Because of symbolic language, individuals were able to discuss with their peers the immediate and long-term damage that bullies—and also cheaters—could do to their own personal interests. They could discuss such problems in private until a powerful group consensus formed, and then they could either openly come out against such behavior by using social pressure and threat of punishment, or they could actively punish or even eliminate serious intimidators who insisted on being active.[64] As a result, many potential bullies (also thieves, cheaters, and other freeloaders) could be routinely held down at the level of phenotype, while the genes of free riders who insisted on being active—those who failed to control their predatory tendencies—were seriously disadvantaged because ostracism, shunning, banishment, or execution could quickly come into play.

In the next chapter, we'll see that with respect to hunter-gatherer capital punishment bullies appear to be singled out much more frequently than deceptive types of free riders like thieves or cheaters; in Chapter 7 we'll see that there was a wide array of lesser sanctions that punished these same bullies but allowed them to reform. This, too, worked to the advantage of altruists because these genetically disposed, would-be free riders (not only bullies but also deceivers) were being "neutralized" at the level of phenotype.

In today's hunter-gatherers, actually a number of factors combine to keep most of these innately predatory tendencies from being expressed. One is simply the ongoing conformist fear of social pressure and active punishment that Durkheim characterized so well for egalitarian bands, for our evolutionary consciences help us to anticipate such consequences and control ourselves. Furthermore, group members respond positively to group rules simply because these rules have been internalized—by anyone save for a psychopath. Obviously, such rule internalization is not enough to eliminate the free-rider problem—but we may assume that it helps substantially.

Thus, the combination of rule internalization and fear of punishment sees to it that most free-riding behavior is being nipped in the bud, in any egalitarian society. Now, think back to the Mbuti meat-cheater, Cephu, and let's assume the man was not a serious psychopath. In spite of having a conscience and an awareness of consequences, he opportunistically broke a rule he was identified with when he thought he could get away with it and the prize was sufficient. In his case, it was delicious, abundant meat combined situationally with the cover of a dense tropical forest, and Cephu went right ahead and cheated—presumably because he thought the risk was very slight. As a means of sorting out useful from personally injurious actions, Cephu's evolutionary conscience was wrong this time, and being aggressively shamed to a point of humiliation by most of his peers would not be forgotten; furthermore, banishment from the band constituted a future threat that could cost him and his kin personal hardship and loss of fitness.

Thus, as Colin Turnbull tells us, this cheating free rider was rendered unlikely to cheat again—and his band didn't have to banish him or kill him to solve the problem. In fact, his genetic fitness remained largely unscathed even though he was, at heart, such an obvious (and arrogant) free rider. Cephu's case is interesting because he was not just a cheater; he also had strong tendencies to aggrandize his own status and behave as an alpha male. The rest of the band made it clear that he could not act on these impulses and remain a member of the band.

Even though there's a large psychological and ethological literature on cheaters and cheater detection, generally free-rider suppression with respect to aggressive bullies has not really been taken into account so far in the basic mathematical models that have anchored the study of human altruism.[65] However, Ernst Fehr's experimental evolutionary economics group in Zurich has discovered that with children participating in experiments in which lucrative offers are made and are either rejected or accepted, there's a tendency to retaliate against those who make selfishly very low offers in order to avoid inequality among the subjects.[66] This "inequality aversion" fits nicely with what I emphasized in *Hierarchy in the Forest,* namely, that human groups have been vigilantly egalitarian for tens of thousands of years because we have inherited tendencies from our ape ancestor to resent being dominated and being placed in a disadvantageously unequal position.[67] In my opinion, further work in this area will be necessary if the important scientific puzzle of human generosity is to be fully addressed.

Generous altruists are vulnerable to cheaters who, in effect, are "designed" to take advantage of altruists.[68] When it comes to bullies, they are designed to take selfish advantage not only of altruists but also of anyone else who cannot or will not stand up to them. The potential effects on human gene pools surely have been substantial. Both selfish bullying and selfish cheating can be considered free-riding propensities that are quite thoroughly suppressible at the level of phenotype, with some help from a multipurpose conscience which, as we shall see, most of the time, for most people, is quite effective at keeping us out of serious trouble socially. When the conscience isn't up to the job, social pressure and then active punishment will phase in.[69]

Here's my evolutionary hypothesis: when bullying is labeled socially as being deviant and is rather thoroughly suppressed at the level of phenotype, the selection agencies we reviewed earlier in this chapter, those that favor altruistic genes but are vulnerable to free riding, can come into play much more strongly. To nullify the potential gains of a would-be or actual bully, there's obviously no detection problem, and if his conscience doesn't restrain him, what it takes to hold him down is a firm resolve of other band members to keep him from asserting himself. They have to stand up to him, and if he doesn't "get it," then the next step for a desperately egalitarian band is to desert him or banish him if possible or do him in as a final solution.[70]

KALAHARI AND INUIT EXAMPLES

For the !Kung Bushmen, anthropologist Polly Wiessner reports that one of the most frequent reasons for talk that involved criticism was "big-shot" behavior, with several dozen cases she collected over several decades of fieldwork. The obvious effect was not only to nip any potentially serious domination attempts in the bud, but also to deter many likely dominators from even making such an initial move.[71] Because selfish bullies cannot readily express themselves, this bodes well for the genes of the more generous or gentler souls who would otherwise be their victims. This benefit holds for the Bushmen today and as we'll see it was true for culturally modern Africans 45,000 years ago.

Anthropologist Jean Briggs gives an account of how such subtle deterrence can have its effect on an Inuit individual (almost always a male) who is "gifted" with unusual propensities to self-aggrandize or selfishly dominate his fellows. The person in question was her adoptive father, Inuttiaq, and in the light of this discussion his personal success at self-control requires some scrutiny. According to Briggs, he had what we might call an exceptionally strong ego and emotionally he was far more intense than his fellows. In her field notes, Briggs's very first description of him used the phrase "barbaric arrogance," and her immediate response was one of fear. This reaction came early in her fieldwork when she needed to become an "adopted daughter" of someone in a small Utku band, and in her non-Eskimo eyes she found this particular "father-candidate" to be atypically unsmiling, hostile, and haughty. She says, "The predominant impression was of a harsh, vigorous, dominant man, highly self-dramatizing."[72]

As for the Utku themselves, Inuttiaq's self-assertion was expressed in ways that were in fact socially acceptable. For instance, he had an unusually aggressive manner of driving his dogs when they were pulling his sled. He also was unusually aggressive in joking with people, but nevertheless, in a society in which aggressive people were felt to be scary, Inuttiaq was basically well thought of because his self-control was so exemplary. Unlike the volatile ethnographer, whose social woes we'll be meeting with in a later chapter, Inuttiaq never lost his temper.

Briggs applied her psychological expertise to this man, and on a troubled note she reports that he outdid other men and women in camp in brutalizing their tethered dogs. He also had violent fantasies that surfaced when he described to Briggs what he would like to do to outsiders (whites) who were more powerful than himself, fantasies that involved stabbing, whipping, and murder. Briggs also notes that if his fellows admired Inuttiaq for never losing his temper, they at the same time feared him for the same reason: "They said that a man who never lost his temper could kill if he became angry; so, I was told, people took care not to cross him, and I had the impression that Allaq, his wife, ran more quickly than other wives to do her husband's bidding."[73]

Was Inuttiaq aware of the tightrope he was walking with a people who were capable of doing away with someone who became unduly intimidating? Briggs thinks this was possible. For instance, Inuttiaq was given to far more aggressive joking than most, his unique specialty being to grab at the penises of younger males. He explained this by saying, "I'm joking; people joke a good deal. People who joke are not frightening."[74] This in a culture in which there were fears that a moody person might stab someone in the back while the two were out fishing and in which, before the Royal Canadian Mounted Police arrived, one man might suddenly slaughter another in order to take his wife.[75]

Seldom does an anthropologist assess an indigenous personality in such detail, so we are fortunate to have this description. The analysis rings true to ethnographic common sense. Apparently, Inuttiaq understood his own unusual dominance tendencies in the light of the strongly egalitarian ethos of those around him, and the multifunctional evolutionary conscience we will be discussing further in Chapter 5 enabled him to continually restrain himself. He was able to achieve this self-restraint because he shared the same internalized social values as his fellows, and because he was shrewd enough to understand when expressing his aggressions (as with his canine victims and when joking) would be acceptable, and when not.

Inuttiaq's situation provides an example of how the threat of social disapproval and group sanctioning can keep a person who by nature is unusually aggressive and dominant so inhibited that if such dispositions might lead in the direction of becoming an opportunistically bullying free rider, they simply remain unenacted. My guess is that if an ascendant social position somehow became culturally acceptable among the egalitarian Utku, Inuttiaq the superficially docile good citizen might well have become more of a leader than he was. And had his social sensitivity and self-control been less, it's conceivable that he might even have riskily tried to become something of a camp bully, in a group that was determined to stay egalitarian. Perhaps his exceptional ego would not have led to this, for his drive to dominance appears to me to have been not very extreme by Eskimo standards.

Unfortunately, I know of no other ethnographer who has provided such a detailed portrait of a man who by nature seems to be unusually assertive. Once in a while a really driven dominant Inuit male is able to play this role of intimidator or even despot for a time.[76] All the Inuit are egalitarian; they favor a humble, generous type of person, and they hate and fear a selfish aggressor who violates their dearly held code of equality. When a man seems to be intent on such serious self-aggrandizement, eventually his peers will find a way to deal with the problem, and if he proves to be incapable of reform, the solution may be a final one.

A FAVORED HYPOTHESIS

Earlier in this chapter we considered a range of possible selection mechanisms that might be able to support extrafamilial generosity.

Three of these evolutionary mechanisms can be greatly empowered by the substantial neutralization of two types of free-riding behavior, which take place when potentially serious self-aggrandizers rein themselves in with the help of their consciences, or when selfish, aggressive cheaters like Cephu are actively put down.

Let's begin with selection taking place at the between-group level. In thinking about how altruistic traits could be sustained by group selection, Richard D. Alexander flirted fairly seriously with this type of explanation, with prehistoric warfare[77] as an enormous and acknowledged wild card—and obviously with the porosity of hunting bands being a drawback because this dilutes the effect.[78] I have shared Alexander's interest in this possibility, and recent work by Bowles[79] has brought a new dimension to what has been a sometimes acerbic group selection debate.[80] Indeed, the case for group selection as a factor in the evolution of altruism is becoming increasingly strong, and the arguments I'll be making in this book about free-rider suppression will make it stronger still.

In my opinion reciprocal altruism must be largely set aside because it applies mainly to dyads, but it does encompass child-rearing partnerships, which usually are dyadic. This could account in part for the altruism found in foraging bands because breeding partners are seldom very closely related, even where cousin marriage is the ideal. And if we assume that most continuing marital relationships involve approximately equalized mutual inputs and benefits, this could be a positive factor in the selection of altruistic traits. However, cheater detection is crucial with respect to female adultery, and in hunting bands group suppression of such behavior is far from consistent and not necessarily very effective.

With respect to selection by reputation, in a small band that is talking about people's behavior all the time, it's far more difficult to dissemble a good, generous reputation than it is in a modern urban society with its relative anonymity. With free riding basically immobilized, selection by reputation—as a distinctively human type of social selection[81]—could be an important and efficient means of favoring extrafamilial generosity. This would hold not only for marriage choices but also for choices of subsistence partners and for choices of political allies in and out of the band, for who is favored or disfavored in extending safety-net help, and more generally for situations in which families are choosing to live in one band or another and must be granted permission to do so.

As a mechanism for selection in favor of self-sacrificial generosity, I believe that social selection needs the most theoretical development. And while no one of these mechanisms we have discussed could have done the job alone, I suspect that social selection will prove to be very important. For humans as I shall be defining it, social selection involves a unique combination of selection by reputation and free-rider suppression, and, as we'll be seeing later, by itself reputational selection contributes to powerful interactive effects similar to those found in Darwinian sexual selection. There, exaggerated maladaptive traits like peacocks' resplendent but unwieldy tails are kept in place by female choice, which at the level of gene selection serves as a means of compensation.

Altruism, too, is by definition basically maladaptive, which means that unless group selection is strongly operative, some kind of individual compensation must be taking place. Social selection probably could not fix altruistic genes in our species' gene pools all by itself; this will require further research by scholars who do such modeling. But it seems likely to have been a leading force in what was a multifaceted selection process based on contributions from a number of mechanisms,[82] including reciprocal altruism and group selection.

In this book a heavy emphasis will be placed on the two types of social selection we have discussed here, one of which is reputational selection and the other is selection that takes place when groups crack down on deviants. In both cases, we will be exploring the role of human preferences, which stem from human nature, in further shaping that same human nature.

4 Knowing Our Immediate Predecessors

FROM PRESENT TO RECENT PAST

In this and later chapters I'll be looking for predictable kinds of anti-free-rider social control found very widely today so that I can confidently project them into the more recent hunter-gatherer evolutionary past, and thereby assess their impact on evolving human gene pools. This treatment actually began in the previous chapters, but here I'll try to justify making such projections. I'll be particularly interested in dire forms of punishment that could have affected individual reproductive success drastically and therefore could have strongly shaped gene frequencies and, ultimately, human nature.

To reliably make the case for the punitive type of social selection's having acted on our genetic makeup, it will be useful to project the group behaviors of today's foragers into the more recent Pleistocene past as conservatively and accurately as possible. This means we must limit our reconstruction to human predecessors who had brains equal to our own and whose cultures had become flexible and advanced like our own. Archaeologists call these people culturally modern humans, and it's widely agreed that in Africa they had arrived by 45,000 BP.[1]

SETTING ASIDE THE "MARGINALIZATION" ARGUMENT

In the African archaeological record, cultural modernity is assessed in terms of a rather abrupt appearance of more complex and regionally changeable stone tool technologies, along with objects of self-adornment and "art," often in the form of engravings. However, interesting as these developments may be, they tell us precious little about what was happening with these people socially. For this reason, it will be necessary to use today's foragers to reconstruct the group life of their predecessors.

Past attempts to do so have met with major objections from scientists who deal in human prehistory, so we'll have to get technical here. For doubters like the influential late political anthropologist Elman Service or more recently archaeologist and hunter-gatherer expert Robert Kelly,[2] a main problem is that most of today's foragers have been "marginalized" by aggressive tribal agriculturalists and, eventually, by civilizations and then empires that took over our planet's more desirable areas. In contrast, Pleistocene foragers had their pick of world environments, and, so the theory goes, they didn't have to cope with the not-so-productive semideserts, arctic wastes, and other marginal habitats that often limit subsistence possibilities today. Thus, there's no telling what they were up to.

Service made his persuasive marginalization argument more than three decades ago, and it made sense at the time. Unfortunately, it has become almost a truism in archaeological and evolutionary circles that Pleistocene foragers must have been living in a fat city situation because unmarginalized small populations could pick their rich environments at will. However, the available prehistoric information has changed since then, and changed dramatically. What's new is our understanding of Late Pleistocene climates and their all but unbelievable instability.[3] Frequently, and cyclically, rapidly changeable weather patterns could have led to two kinds of prehistoric "marginalization" that, roughly speaking, would have been comparable to what we see today.

One was purely ecological marginalization. This was likely when areas with adequate patterns of rainfall became drier and only smaller populations could be supported in widely scattered bands. Such climatic downturns could have created localized-drought challenges directly comparable to those arising in the capricious Kalahari Desert area that people like today's !Kung and !Ko Bushmen have to cope with, or challenges faced by desert Aborigines in Australia or Great Basin foragers in North America.[4] The second type of marginalization would have been political. As cyclically better conditions allowed Pleistocene groups to multiply, competition could have intensified as more aggressive hunter-gatherer groups began to monopolize the better resources, marginalizing other foragers just as today's foragers have been marginalized by territorially aggressive farmers.

Not only that, but when there were shifts toward ecological good times that permitted gradual but eventually substantial population growth, and then a sudden downturn arrived, it's likely that foragers of one language or ethnic group would have been prone to aggressively push aside foragers of another. This would have been especially the case if the resources they were competing over were rich enough, and concentrated enough, to be readily defensible.[5] Such marginalization could have resulted in outright warfare, and even though direct prehistoric evidence before 15,000 BP is lacking, to judge from certain foragers today[6] such conflict could have become quite intensive under the right conditions.[7]

Some of the Holocene microclimates that today's foraging nomads deal with are not just thin in resources but are quite unpredictable in the shorter term, and at least a few cases of famine have been recorded by ethnographers.[8] In the recent Holocene these perilous junctures have occurred so rarely that only occasionally does an anthropologist even visit a field site at which a true famine is well remembered. But a striking exception is certain Inuit speakers like the Netsilik in central

Canada or Inuit groups in Greenland,[9] while Bushmen and many other foragers living on semideserts at least are able to recall episodes of serious privation.[10] The social, emotional, and genetic effects of such dire scarcity will be weighed in Chapter 10.

Using today's nomadic hunter-gatherers as models for their nomadic predecessors must be further justified here because Service's "marginalization taboo" still enjoys such wide adherence. Of course, when archaeologists show their unreadiness to reconstruct the social life of "prehistoric foragers,"[11] often this involves a legitimate fear of projecting modern human behaviors on to much earlier types of humans who had smaller brains and, in all likelihood, had a significantly different behavioral potential. With respect to smaller-brained humans who had not yet developed culturally modern tool kits, such conservatism has been and still is quite appropriate. Here, however, I am considering only the more recent prehistoric humans who matched us in brains and cultural capacity.

My theory is that the main outlines of their social and ecological life can be reconstructed quite straightforwardly simply by identifying behavior patterns that similarly nomadic foragers share very strongly today. However, such reconstructions must be carefully strategized, and I will be reconstructing only what might be called core behavior patterns,[12] that is, behaviors involved with gaining a living, along with the social behaviors that are basic to such an enterprise. In addition, in recreating Late Pleistocene socioecology, I will be focusing only on those carefully selected contemporary foragers whose ecological lifestyles would have been likely 45,000 years ago.

FINDING THE RIGHT HUNTER-GATHERERS

This analysis has involved ten years of research effort.[13] My first task was to evaluate the great majority of the world's ethnographically described hunter-gatherer societies, 339 of them,[14] to weed out those that obviously would have been atypical in the Late Pleistocene. I eliminated, for instance, the many North American mounted hunters like the Apache or the Comanche because horses were domesticated only recently.[15] I also eliminated a few bands that lived dependently at missions, like the well-known South American Aché, and ones that symbiotically traded food with horticulturalists like the Pygmies or the Agta of the Philippines or foragers who had begun to cultivate a few plants themselves. And then I had to set aside dozens of societies that had been heavily involved for centuries with the European fur trade, such as the North American Ojibwa and the Cree, and of course I had to eliminate several dozen sedentary foraging societies that began to intensively store food and eventually lost their egalitarian ways to become markedly hierarchical, like Japan's aboriginal Ainu or the Kwakiutl of British Columbia—who actually had slaves. After this triage was finished, only about half of the world's foraging societies were left. They were uniformly independent, nomadic, and egalitarian, and they were suitable—if used in quantity with some statistical sophistication—as models for humans in the latter part of the Late Pleistocene Epoch, which overall lasted from about 125,000 BP until our present Holocene Epoch began to kick in.

The contemporary models I'll be using, then, are taken from the perhaps 150 groups that I'll be referring to as "Late Pleistocene appropriate"[16] foraging societies, or, in a more streamlined fashion, as "LPA foragers." My assumption is that they are very similar to the culturally modern people who were evolving in Africa around 45,000 BP and were spreading to most parts of the world.[17] (Keep in mind that the people who painted those beautiful cave paintings in Spain and France first evolved their artistic potential in Africa, where cultural modernity had its beginnings.)[18]

With a third of these worldwide LPA societies now coded in fine detail with respect to their social life, this is what I've found so far. To start with, these fifty societies are definitely all mobile, and as nomads, instead of trying to store their large-game meat as individual families, they share it widely. It doesn't matter whether these people live on Arctic tundras or in tropical forests—they never dwell in permanent, year-round villages, and they always combine hunting and gathering to make a living according to what is environmentally available, with an emphasis on eating the relatively fatty meat of large mammals. Normally, their camps or "bands" average around twenty to thirty persons, and each family cooks at its own hearth.[19]

In the case of camp size and the butchering of large game, today's ethnography coincides with what we know of yesterday's archaeology.[20] Just from the ethnography, we know that invariably these people believe in sharing their large game with everyone in the band, and that they all face problems of social deviance like bullying and theft and employ similar basic means of social control to combat them. These foragers very predictably share a core of moral beliefs with an egalitarian emphasis on every hunter's being a political equal, while the political positions of women as nonhunters are much more subject to diversity. We also know that their bands involve highly flexible camping arrangements, with families moving in and out as needed, and that at any given time a band will be composed of a mixture of some related and many more unrelated families.[21]

If these bands were just big extended families, the cooperation and altruism they engage in would be much easier to explain, for kin selection theory would do the trick. But they aren't, and we may readily assume that the same was true 45,000 years ago. That's why, as our evolutionary story unfolds, we'll be so interested in seeing by proxy how prehistoric forager lifestyles could have generated distinctive types of social selection, as agencies that could have supported generosity outside of the family at the level of genes.

Today, the social patterns I've discussed hold all but uniformly across an almost incredibly wide variety of environmental niches that these LPA foragers manage to cope with successfully. These range from arctic tundras to boreal forests in the far north, to productive temperate or tropical forests, to resource-stingy jungles, and to fertile plains or game-rich savannas and arid semideserts.[22] These environments include coastal areas as well, which prehistorically were likely to have served as refuges from glacial cold snaps or droughts. These sites today often would be under water, and it's conceivable that people could have become sedentary for a time while exploiting them. It's even possible that sometimes they did so for long enough to begin to lose their egalitarian, meat-sharing lifestyle if a long-term habitat was rich enough to permit food storage. However, while families' economic standards of living may have begun to differ, it's likely that *political* egalitarianism would have been more resistant to change, and in any event these outliers would not have negated the social central tendencies I've been describing; they would have held very widely.

Climates today range from hot to frigid and from stable to sometimes fairly unpredictable, but before the Holocene phased in, Late-Pleistocene-type climates could change with a rapidity we seldom see today. It's no accident that during the lengthy Pleistocene Epoch, human brains just kept on getting bigger, for we've had a lot of challenges to cope with,[23] and surely some of them involved situations of desperation and famine. In Chapter 10 we'll learn, from today's foragers, exactly how desperate these situations were likely to have been and what could have happened to the usual food-sharing practices when people were facing actual starvation.

It's remarkable that a single main "type" of band composition and group life can work so successfully when such a startling array of environmental challenges is faced, but this in fact is the case. Scholars agree that socioecological flexibility is what makes this possible, and although the band is an obvious focus, to get the total picture we have to think in terms of many culturally similar bands dispersed over sizable regions, with families changing bands on a rather frequent basis. In the Late Pleistocene with its dangerously capricious environments, very likely this highly flexible approach to group living and subsistence was not merely convenient but often absolutely necessary to getting by—with the sheer survival of entire bands or regional populations surely being on the line much more frequently than is the case today.

In that epoch, as today, there would have been at least a few exceptions to the basic overall patterns—that is, the strong central tendencies I've been describing. I've just suggested that temporary sedentary adaptations were likely, and the food storage could have reduced the sharing of large game. Another readily understood contemporary exception that was likely prehistorically can be seen in the few foraging societies that cope with environments so spare that most of the time they are able to forage only as families, without forming bands—as with some desert Australian Aborigines, who subsist partly on insects, or as with certain of the Shoshonean Indians living in America's semidesertic Great Basin area, whose fat and protein come mainly from fluctuating harvests of piñon nuts rather than from wild game.[24] In the unstable and periodically dangerous Late Pleistocene, occasional divergences from the central tendencies I described previously were likely to have been more frequent. However, the great majority of these prehistoric foragers would still have followed today's main pattern, meaning that they lived in mobile, flexible, egalitarian multifamily bands of twenty to thirty and they invariably shared their beloved large game with its exceptional fat content. That, I propose, was the central tendency for those large-game hunters, and surely it was a strong one, then as now, even though these and possibly some other outliers were likely.

I've taken 45,000 years before the present as the time when *Homo sapiens* populations in Africa had become culturally modern; this means that they had a full capacity to flexibly invent and maintain the remarkably variable material and social patterns that LPA foragers exhibit today. However, this date may be somewhat conservative,[25] for humans had already become *anatomically* modern by 200,000 BP,[26] which means that they were then at least physically indistinguishable from us. Increasingly, it's looking as though *cultural* modernity, as deduced from the making of increasingly complex and variable artifacts, some of them symbolic, was phasing in earlier than 45,000 BP. The problem is that cultural modernity evolved in Africa and African archaeology is just getting up a real head of steam. Thus, even though I shall use the 45,000 BP figure to keep the analysis conservative, we might put in its place a date of 50,000 BP or even 75,000 BP or earlier. Only time, and more excavations, will tell.

LETHAL SOCIAL CONTROL AT 45,000 BP

Much of this book is about how punitive social control, a harsh form of social selection indeed, has acted on human gene pools. I shall be proposing that aggressive (and originally nonmoral) social sanctioning shaped the earlier human genome to give us an evolutionary conscience, and that extensive curbing of free riders was another important effect. In turn, free-rider suppression opened the way for the evolution of altruism; this development will be explained in detail in chapters to come. These three developments, taken together, may be seen as the scientific story of moral origins.

Both punitive and positive social selection were closely involved with group political dynamics, and when band members started to form consensual moral opinions, and were systematically punishing deviant behaviors and rewarding prosocial behaviors, a novel and powerful new element was added to human evolutionary process. The ultimate result was the human nature we carry around with us today, which of course combines selfish egoism with nepotism but also includes enough sympathetic altruism to make a major difference socially.

The more Draconian the sanctioning actions of angry groups were, the stronger the force of punitive social selection would have been as it acted on prehistoric gene pools. Lethal attacks on disliked individuals by sizable coalitions can be projected back into the Late Pleistocene Epoch with great confidence, for as we'll see in the next chapter, such killings had at least a significant precursor in the shared ancestor of humans and two of the African great apes. However, among today's LPA hunter-gatherers it's difficult to say, for at least two reasons, exactly how often they inflict capital punishment on one of their own. First, when an ethnographer visits a foraging society for a year or two, it's extremely unlikely that a deviant's being killed by the band will be witnessed or even be talked about—unless the right questions about remembered history are asked. Second, band executions that have taken place say a hundred years before or earlier may simply be lost to memory.

Here we'll be surveying my entire coded sample of fifty LPA societies to get at least a partial idea of how often capital punishment has been taking place over the past century or so. But keep in mind that we'll be seeing only the tip of an iceberg, not only because of the two reasons I just gave but also because most hunter-gatherer ethnographies, although obviously precious, are at the same time likely to be seriously incomplete with respect to capital punishment because indigenous people quickly learn that missionaries and colonial administrators view their executions of dangerous deviants as "murder." As a result, they quickly learn to clam up, and often stop this practice.

Table I Capital Punishment in Fifty LPA Foraging Societies*

Type of Deviance	Specific Deviances	Societies Reporting
Intimidation of group	Intimidation through malicious sorcery	11
	Repeated murder	5
	Action otherwise as tyrant	3
	Psychotic aggression	2
Cunning deviance	Theft	1
	Cheating (meat-sharing context)	1
Sexual transgression	Incest	3
	Adultery	2
	Premarital sex	1
Miscellaneous	Violation of taboo (endangering group)	5
	Betrayal of group to outsiders	2
	"Serious" or "shocking" transgression	2
Deviance unspecified		7
Total societies reporting capital punishment		24

*The above figures are derived from the author's hunter-gather database.

Table I shows the basic patterns of active capital punishment (almost all of males, by males) that did emerge in spite of such seriously incomplete data. I should mention here that just a few bands had a formal system for deciding to eliminate a deviant, with a council of elders meeting to come up with a death sentence. More often the process was far less "structured," which means that the entire band (including the females) would simply agree, informally, that a close relative of the deviant should do him in. The deviant was almost always a "he," and there was a very good reason for his executioner, a well-armed hunter, being a close relative. This will be explained in Chapter 7.

This intensive survey of fifty LPA hunter-gatherers involved coding over two hundred ethnographic sources, and half of the time, surely for reasons that have just been explained, capital punishment wasn't even mentioned. For almost half the societies, however, some instance of capital punishment was in fact reported, as seen in Table I, and often the ethnographers were in a position to specify the deviance patterns that brought it on.

Eventually, I will have coded at least three times as many of these societies, which will make such statistics less subject to the vagaries of sampling error. But these numbers do enable us to pick out one very salient pattern: half the people killed, almost all of them males, were *intimidating* their groups. This was done by greedily or maliciously using supernatural power to seriously threaten the welfare or lives of others; by being far too ready to kill, repeatedly, out of greed or anger; by otherwise managing to seriously dominate others; or (much more rarely) by being aggressively insane.

This main pattern fits with the previously emphasized fact that all LPA bands today are highly egalitarian in their social worldview, which means that groups quickly become angered when individuals given to domination (ones like the Pygmy Cephu) try to self-aggrandize. Except for the killing of psychotically aggressive men, all of these aggressive intimidators were considered to be *morally* deviant, and other research I have done, including that set forth in my book *Hierarchy in the*

Forest,[27] suggests that if small nonliterate groups are to keep in place the egalitarian political orders they so strongly prefer, they sometimes will be obliged to use capital punishment when a menacingly aggressive personality appears in their midst.

The choice is between suffering a determined tyrant or taking him out, and it's precisely because people inherit varying dominance tendencies from their parents that in any small egalitarian society, eventually someone with proclivities to behave far too dominantly will show up, act on these propensities, and manage to get himself in serious trouble. This is either because he doesn't have the ability to properly read the situation or because he lacks the capacity to restrain himself. The gender pattern is as follows: it is almost always a well-armed male hunter who tries to dominate his peers, but when a band coalesces to bring down such a tyrant, the females may be as active as the males in the political dynamics involved, and in one rare execution that was actively communal, they became physically involved.[28]

If we look now at the several low-frequency effects in Table I, we see that stealthily breaking rules to take advantage of others by cheating or theft are at least mentioned. With respect to free riding, this means that both intimidating and deceptive free riders can receive a death sentence, but keep in mind, here, that we are considering only reasons for *capital* punishment. Lesser sanctions, like ostracism, shaming, and expulsion from the group, are more likely for lesser crimes that do not necessarily threaten an entire group's safety, autonomy, or welfare. In Chapter 7 I will present data on these lesser sanctions, which do not suffer so much from the incomplete data problems I have described with respect to capital punishment.

Being punished capitally will have varying but always deleterious effects on the reproductive success of the serious malefactors who are killed. This obviously precludes having further offspring, so if a man is executed while in his twenties or thirties, the fitness consequences are enormous. Furthermore, after his loss the offspring he's already created will receive less parental support and their fitness (half of which is shared with the father) will suffer accordingly. Furthermore, he will not be around to help coresident closer kinsmen like siblings if they need it or to cooperate with them.

PUNITIVE SOCIAL SELECTION

As human gene pools evolved over evolutionary time, the effects of active group punishment on individual reproductive success could have been quite significant, and surely these widespread effects became increasingly influential as people gained stronger consciences and their group punishment increasingly became driven by moral outrage. Here we're speaking not only about capital punishment but also about banishment and serious ostracism, which can interfere significantly with an individual's gaining the benefits of cooperation. In addition, just acquiring a bad moral reputation surely would have made some people avoid such deviants in important choice situations such as marriage or in looking for subsistence partners.

All of these mechanisms entailed *social selection* in the sense that preferences shared by groups were affecting gene pools.[29] More specifically, all involved *negative* preferences, and all disadvantaged the reproductive prospects of individuals prone to social deviance—or at least those who could not control their sexuality, greed, or inappropriate hunger for power. For such moralistic social selection to have been a significant factor in shaping human gene pools, probably it had to be operating for at least a thousand generations,[30] which in long-lived humans comes to about 25,000 years. Thus, the period just before cultural modernity arrived could have figured importantly in moral origins—if this development took place at all quickly.

In the next chapter we will start at the beginning to consider in more detail the head starts that were provided by two distant but highly relevant ape ancestors, the Common Ancestor and Ancestral *Pan*, so as to discern how some special developments in the human line led us, but not chimpanzees or bonobos, to evolve a capacity for virtue—and shame.

5 Resurrecting Some Venerable Ancestors

GETTING TO KNOW OUR NEXT OF KIN

Moral evolution can be defined most basically in the light of our uniquely human conscience functions and how they were able to evolve. But to fully understand how we acquired a shameful evolutionary conscience, we must go back millions of years to the ape ancestor that provided us with important building blocks,[1] out of which somehow the moral life of humans was "constructed."

Today, we're beautifully poised to reconstruct many of these ancestral behavior patterns, but yesterday we weren't. As recently as the 1980s, physical anthropologists around the world had differing and uncertain thoughts and theories about who our closest living primate relative might have been. For some, chimpanzees in Africa had something of an edge over that continent's gorillas, whereas for most the Asian orangutan trailed far behind and the lesser apes and monkeys were pretty much disqualified.

This was sheer guesswork, but we could at least assume that some ancestor must have been knuckle-walking across the floors of African tropical forests and climbing tall trees to eat large quantities of fruit, and it made sense that this had to have been taking place *tens* of millions of years ago because we humans were so different from other primates in our cognitive abilities and our complex social life. That's what everybody thought, and in much of this they were far from being correct.

Several decades after James Watson and Francis Crick had unraveled the basic mystery of DNA,[2] laboratory geneticists were finally in a position to compare the genomes of humans and those of the three African great apes considered the most likely candidates for being our closest "cousins." The genetic findings astonished the scientific world.[3] It turned out that humans, chimpanzees (*Pan troglodytes*), and also a slightly smaller African "chimpanzee" called the bonobo (*Pan paniscus*) all shared over 98 percent of their DNA—a far greater similarity than anyone had imagined. And gorillas were not far behind, even though humans and the two *Pan* species turned out to be even more similar than *Pan* and gorillas were. It followed that all four of us shared an ancestor more recent than anyone had suspected.

Such reckoning is done through a "molecular clock" analysis, and the estimates are reasonably precise because gene mutations accumulate at a fairly predictable rate and can be counted. Once one species branches into two, over time their genomes will increasingly differ, and when the genetic makeups of humans (*Homo*) and the pair of *Pan* species are compared, the finding is that these two lineages diverged merely about 5 to 7 million years ago (MYA). For purposes of streamlining, we'll get rid of the "plus or minus" and call it 6 million years. We've had this figure available for two decades now, and an evolutionary time gap of merely 6 million years makes for a very close relative indeed—even though "Ancestral *Pan*" surely looked far more like an ape than a human.[4]

It was primatologist Richard Wrangham who in 1987 first identified and began to socially describe a somewhat older predecessor he called

Figure I Phylogenetics of Human Ancestry

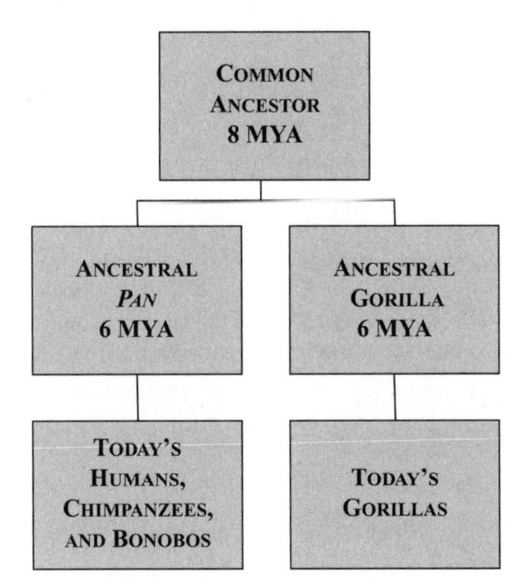

the "Common Ancestor."[5] This CA is the predecessor that the two *Pan* species and *Homo* share with gorillas (see Figure I), and in this case the molecular clock gives us an estimate of about 7 to 9 million years, which we'll round off to 8 million years. With our two direct ape ancestors identified and placed in time through DNA analysis, we can begin to systematically reconstruct some (but not all) of their social behaviors—something that was all but unthinkable before Watson and Crick.

This will be of critical importance to the analysis to come. First, it will enable us to identify ancient adaptations that fortuitously made it easier for the descending human lineage to develop a protoconscience. And building on this, with an increasingly developed sense of right and wrong, we can explain not only the origination of a fully moral way of life, but also the rise of selection forces that allowed us to become as altruistic as we are today. These developments were replete with feelings of shame and blushing, and as we know them today they involve an intensely moralistic interest in the behavior of other people. We just can't help judging people, with shame and virtue on our minds.

If Ancestral *Pan* provided some behavioral building blocks, what about the physical nature of these ancient apes? We have yet to discover a skull or skeleton, or even a handful of teeth and bone fragments, for our immediate prehistoric predecessors inhabited humid equatorial zones where living things decayed or decomposed so quickly that fossilization very rarely took place. However, we are fortunate in that many important *behavioral* characteristics of these ancestors can be reconstructed, quite reliably, without our ever seeing a tooth or measuring a cranium. This is accomplished by looking for major behaviors that are shared unanimously by all the living descendants of the ancestor in question, and the methodology is called "behavioral phylogenetics."[6]

This enables us to reconstruct an ancestor's behaviors if and only if such behaviors remain present in *all* of its living descendants. It may seem remarkable that we can reliably discern behavior patterns so distant in time, but the conservative processes of natural selection are prone to follow a relatively straightforward and direct path as they keep a species in business over evolutionary time. Biologists call this "parsimony," and they call these original, ancestral traits "primitive."[7] Thus, if all four descendant species tend to eat lots of plant foods, which they do, it's parsimonious to assume that this primitive trait came straight down from the Common Ancestor—and that it continued to be useful in the changing environments that were experienced subsequently, or else it would have simply withered away.

In the social sphere, Wrangham identified several very basic social constants. He determined that all four of the Common Ancestor's descendants live today in social groups with definite boundaries from other groups, so this pattern had to be primitive, meaning ancestral, and therefore today it is homologous, which means it is based on similar genes.[8] Another homologous commonality was stalking and attacking other members of the same species, so we may assume that the CA was doing this at 8 MYA and that all four descendant species have been doing so ever since. In *Demonic Males,*[9] which explores the roots of human violence, as a theorist Wrangham also considered the more recent ancestor that I've referred to as Ancestral *Pan*.[10] Its primitive behaviors will figure importantly in our determining how a conscience came to evolve, and presently we'll be reconstructing these behaviors in some detail.

I must quickly mention that our main interest will be in Ancestral *Pan*, an ape that basically carried on with all the behaviors we can identify in the CA and added some more. As we'll be seeing, once gorillas are removed from the ancestral equation—and this divergence took place about 6 MYA—some very significant behaviors can be added to Ancestral *Pan*'s reconstructible behaviors. But first I wish to extol the technical virtues of using homologous reasoning in evolutionary behavioral reconstructions.

THE ADVANTAGES OF TALKING ABOUT SHARED GENES

Using a natural-historical approach requires distinguishing carefully between behavioral similarities based on homology, which means shared genes, and those based upon analogy, which merely means similar functions. It's important to differentiate between the two because in a pair of species that shares an extremely remote ancestor, totally different mechanisms may underlie superficially similar-looking behaviors that have converged. For example, when a territorial Norway rat instinctively attacks a nongroup member, this rat is programmed to react aggressively to any member of the same species that does not carry the odor of its own social community.[11] It's that simple. Even though chimpanzees also attack their neighbors on contact, they have far more "sophisticated" means of identifying them psychologically because they know where their territories are, and they know by sight that their neighbors are strangers.

What we have here is merely a case of analogy: a pair of similar-seeming behaviors, both making for strong group territoriality, are based on some very different underlying mechanisms and hence on disparate genes. This is a matter of convergent evolution, one of the marvels of natural selection. In my dry garden in Santa Fe, New Mexico, I remember vividly thinking that I was watching a new kind of hummingbird one day, as the needle-nosed creature flitted from flower to flower with the sometimes stationary poise of a helicopter, its wings whirring, when suddenly I realized that I was looking at a *moth* and that this hovering insect must have followed its own path in developing genes very different from those of a hummingbird, which nonetheless enabled it to sip nectar while stationary in midair.

When humans attack and kill their enemies, we may assume that the similarities in how we and chimpanzees express hostility toward out-groupers are *homologous,* because we share such a recent ancestor and over 98 percent of our DNA. To be precise, the territoriality of chimpanzees and humans is very likely based on similar mechanisms of psychology, which in turn are prepared by large numbers of genes that are similar. The same would go for bonobos, even though their potential for similarly territorial behavior seems to be far less violent.[12]

WHAT IS A PREADAPTATION?

The moral origins theory I'm developing here is based squarely on arguments of homology. That's why I'll be taking such pains to identify likely primitive building-block behaviors in the CA and also in Ancestral *Pan,* and I will endeavor to show how, over evolutionary time, such behaviors have provided an important preadaptive wherewithal for further evolution.

Vocabulary-wise, preadaptation and the more recent term "exaptation" mean the same thing—genes that support an already evolved trait are "made use of" by blind selection processes in future adaptations. I prefer the older term, even though for many scholars it rings of improper teleological thinking because somehow "preadaptation" sounds forward-looking. Here there will be no such suggestion. Basically, evolutionary process is totally blind—even though through their preferences it has allowed purposeful human actors to unwittingly influence certain aspects of natural selection process in certain prosocial directions. This prehistoric side effect is nothing like genetic engineering by the way, which in contrast is far from unwitting.

With respect to homology, ancestral dominance and fear-based responses to dominance—along with an ancestral capacity to resentfully gang up against dominants—are just two of the many preadaptations we'll be working with. They provided the basis for humans to develop an increasingly potent type of punitive social selection that affected earlier gene pools. At first, this worked mainly through gang attacks, but eventually it led to far more refined means of social control.

The coming analysis of moral evolution will rely not only on a homology-based behavioral reconstruction of Ancestral *Pan,* and on archaeological findings about earlier human evolution, but also on dramatic recent information about Late Pleistocene climates. In that context, there will be some innovation in using current (LPA) ethnographic information to help in reconstructing the earlier evolutionary trajectory of the increasingly modern people who had to cope with these climates. I will show how the moral flexibility we experience today could have helped us to cope with these often unruly and sometimes treacherous climates, mainly by people being able to cooperate—but at times by their being able to set cooperation aside.

ANCESTRAL HIERARCHIES AND ANTIHIERARCHICAL BEHAVIOR

One contribution I have made to the behavioral reconstruction of Wrangham's original Common Ancestor has been to emphasize that it lived a social life that was heavily determined by individual tendencies to dominate and, ambivalently, submit. In *Hierarchy in the Forest* I start by suggesting that if we look closely at gorillas, bonobos, chimpanzees, and humans, we'll see a noteworthy shared tendency for alpha males to appear at the tops of pecking orders, and, linked to the predictable and strong competition for high rank that goes with this, we'll also see that generally subordinates do not relish being dominated. In fact, in all four of these living apes, rebellious subordinates can form counterdominant coalitions[13] to actively reduce the power of alphas. Gorillas do this very rarely, bonobos and chimpanzees can do it routinely, and LPA human groups do this with a real vengeance even though some later types of human groups were able to strongly embrace hierarchy.

Politically, humans are far more flexible than the other three African-based apes. We only have to think of a modern Hitler or Mao or Stalin as an alpha male at the top of a hierarchy—or of an American president as a less all-powerful alpha who nonetheless gives orders to a huge military—and it's obvious that we share the apes' hierarchical tendencies and have a strong potential to develop alpha males. However, the proper humans for evolutionary analysis are the LPA hunter-gatherers I introduced in the previous chapter, and their political arrangements are quite different—even though obviously they are based on the same innate political potential. These humans, along with many of the tribal agricultural people that followed them in time, were strongly and insistently *egalitarian.*[14]

If the other three African great apes can partially neutralize alpha power by forming subordinate coalitions, human foragers have carried such counterdomination virtually to perfection—at least among the adult males.[15] When I say that hunting bands are politically egalitarian, I mean people are so intolerant of alpha-type power moves that normally no individual dares to boastfully aggrandize his own status, let alone try to boss around another hunter or take over a carcass the group wants to treat as common property.

As we'll be seeing in detail, self-aggrandizing individuals who try such moves are put down by criticism or shaming or are banished from the group; and as we've seen already, the very rare unrestrained tyrants, who somehow manage to gain a firmly dominant hold on group life, are soon taken out by capital punishment. Thus, in human politics primitive ancestral tendencies that favor both hierarchical behavior and counterdominant behavior can be expressed phenotypically either in the form of rampant totalitarianism or in the form of radical hunter-gatherer democracy—and anywhere in between. It all depends on how people feel about hierarchy, on how badly centralized command and control are needed, and on the degree to which subordinates' control of those above them can become decisive.

CULTURAL LEARNING SKILLS ARE ANCIENT

Using Wrangham's conservative reconstruction method is quite straightforward. If all of the ancestor's living descendants share a trait, in all probability that trait has to be ancestral. If one or more of the descendants fails to share a trait, however, then an ancestral question mark must be assigned for that particular trait. By using this conservative methodology, even though just a few species are involved, we'll be able to probabilistically peer deep inside the psyche of our two ancestors, to understand much of what made them tick as highly social beings.

Let's begin with "family life" in the CA. We might play to a human bias and ask if in the four extant species there's pair bonding between mating partners, with fathers participating in the nurturing of their own offspring. This certainly is common among humans today and it exists to a significant degree among gorillas, but unanimity is lacking: chimpanzees and bonobos mate promiscuously, and there's no substantial sign that the biological fathers (they can be identified by doing a DNA analysis on their hair samples) show any very special interest in the females they have impregnated—let alone that they take any major care of their own offspring. Thus, as of 8 MYA, two-parent families with strong paternal participation in childcare will have to receive a big question mark.

What we *can* say, however, is that the CA at least had *maternally based families,* for in all four living species the mothers associate very closely with their offspring for up to half a dozen years,[16] while siblings also are socially close. This means that a strong matricentric family pattern goes back to 8 MYA and that minimally this more limited form of familial organization has been present in the line leading to humans ever since the time of the CA. At some point, however, humans obviously added something important: we evolved a two-parent family unit, and even though there are no really solid archaeological cues as to when this occurred, we have only to look at ourselves today to know this did take place. At latest the human nuclear family would have arrived with cultural modernity, no later than 45,000 BP, simply because archaeologists agree so unanimously that we were entirely modern by then.[17]

I chose to discuss this maternally based family structure up front because even such rudimentary familial behaviors provided a very important head start for the evolution of a conscience. Humans are moral because we are genetically set up to be that way, and it's of interest that today's small children turn into increasingly moral beings in highly predictable stages. Sympathetic feelings for others in need of help arrive quite early in infants, as does a primitive sense of right and wrong. These developments are followed by a general sense of rules, while highly self-conscious shame reactions, which include blushing when in the wrong, arrive later when children are reaching school age.[18] These learning windows make up a hard-wired sequence, but the rules themselves are not nearly so explicitly prepared by our genes. Like infant or juvenile apes, young humans learn the social rules of their groups early in life largely from significant others they are with constantly—and the most significant other of all is going to be their mother.

When people learn exactly how to be moral from their cultural environments, this starts with the family. But later in life group traditions can have a very strong or even a definitive role to play, and sometimes cultural tradition can make a given behavior perfectly acceptable in one culture but taboo in another. For instance, in some societies it's abhorrent ever to eat human flesh, whereas in others it may be morally proper and even quite laudable if this act is performed as part of an important ritual or ceremony, as with funerals in the Trobriand Islands of Melanesia.[19] These cultural differences seem to be much more subject to diversity in symbolic humans than they are in the great apes.

CULTURAL DIVERSITY

Thus, the specific rules we internalize from parents and peers are culturally maintained by local groups, rather than being firmly encoded in our genes. However, although many of these rules can vary widely across cultures, among LPA foragers many of the more socially important rules do appear to be universal. For instance, no hunter-gatherer community condones either killing another group member without proper cause, or theft or cheating within this primary group,[20] and the same appears to be true of all humans in all walks of social and cultural life.

Whether a social behavior is unique to one forager group or universal to all humans, our rules of conduct are always learned through cultural transmission—which means originally in the bosom of the family. If we leave morality aside and think ancestrally, all four of the living species we're concerned with are capable of cultural transmission of social rules and behaviors. This begins with *mothers* being closely bonded to their infants for a number of years, with the youngsters engaged in imitation and emulation. Thus, for instance, the art of political alliancing is learned first by an infant's watching Mom's behaviors, then by trying to replicate them, and later by a juvenile's watching adults in the group and further modeling its behavior on theirs.

That's why, for example, these four living species, and therefore the Common Ancestor they share, are all capable of ganging up in coalitions that try to limit alpha power.[21] This same social learning capacity would have continued to be available millions of years later, when archaic humans began to develop a culture that carried a still more decisive message: dominating your peers is not merely irritating to others; it's *morally wrong.*

For moral evolution to have been set in motion, more was needed than a preexisting capacity for cultural transmission.[22] It would have helped if there were already in place a good capacity to strategize about social behavior and to calculate how to act appropriately in social situations—especially when a behavior could get an individual in serious trouble with a dominant other, be this a single individual or, at times, a sizable coalition.

Just like ourselves, chimpanzees and bonobos are quite adept at coping with the power of others, so the development of a rule-based moral capacity had a nice head start. But a head start is only a head start. Next, the right selection pressures had to come along to lead to strong and consistent social control by entire groups, and for these two *Pan* species they apparently didn't come along. For humans they did, however, and the reasons for this will be discussed when we turn to the evolution of a conscience.

HOW SOPHISTICATED WERE ANCESTRAL BRAINS?

As Darwin noted, our consciences guide our actions in ways that help us as social beings.[23] They keep us congruent with the standards of our groups, and therefore they keep us out of trouble with our potentially very punitive peers. Today, we know that our brain's capacity for engaging in social planning and for dealing with complex moral orders is at least partly localized in our quite sizable prefrontal cortex.[24] As we've seen, this brain region plays a major role in guiding our social interactions, and it also involves our moral feelings. In short, it helps us to adjust to the moralistic groups we live in.

A broader question we must address now is exactly how much of an ancestral head start existed in this direction? Is it possible that our Common Ancestor actually possessed something like a primitive but internalized, self-judgmental sense of morals? If we consider all four of the CA's descendant species as a single phylogenetic family, or "clade," we know that in this clade the prefrontal cortex is substantially larger in proportion to body size than that of, say, a rat or a solitary polar bear or even a highly sociable wolf or dog. There was a time when it was assumed that in this respect the human prefrontal cortex had to be uniquely enormous, but careful scientific measurement has shown otherwise.[25] Ours may not be larger than that of the three other African great apes, and this is not terribly surprising given the social sophistication of these apes. In fact, chimpanzees, bonobos, and gorillas can exhibit remarkably nuanced social behavior when we observe them in the wild. Given this overall social acumen, it's conceivable that these apes might exhibit *something* like the moral potential we possess—but that this remains latent because of the "might makes right" type of social life they lead as wild animals.

Proportionate *size* probably is not the only brain factor we should be considering. Natural selection could have changed humans' mental functions by restructuring the brain internally, without affecting its overall size, and could have done so in ways that would be difficult to infer from a fossil record that includes hundreds and hundreds of fossilized brain cases but tells us very little about the brains inside because soft tissue is so unlikely to fossilize. This means we must at least ask whether our phylogenetic "cousins," the living African great apes, may have some potential to react as moral beings who possess a sense of shame. If all of them can do this, then we must think in terms of a very strong ancestral preadaptation in the same direction—and consider moral origins as being not strictly a human affair.

If our Common Ancestor or Ancestral *Pan* had had some self-judgmental reactions by which it felt badly about itself because of rule breaking, this could have been a preadaptation that made it much easier for the human shame reactions we know so well today to evolve. Required also would have been a sense of "self," even though as a single factor this does not guarantee having the capacity for shame.

TESTING SELFHOOD IN THE LABORATORY

Our human brains do provide us with a well-developed sense of self. And every person possesses this type of awareness. We share it with just a handful of other large-brained and highly social species, such as great apes or elephants or dolphins,[26] and there are convincing experiments to back up this statement. In humans what sociologist George Herbert Mead called the "social self" has been extensively studied by social psychologists,[27] and its fundamental role in our makeup is reflected in our languages—which consistently find some way of saying "I" and "you," as well as "we" and "they." As actors on a social stage, we are keenly aware of ourselves—of our *selves*—and this also enables us to realize that others have similar selves, of which we become intuitively aware during social interaction.

At the base of selfhood lies the simple capacity for visual self-recognition, which all people possess. For example, we routinely look at ourselves in the mirror and are aware that it is not another individual but our own image that we are seeing there. Even though humans obviously did not evolve with mirrors in hand, our large brains and our natural history as highly social beings provide the foundations for self-recognition. This trait is so well established that if a naïve member of a nonliterate culture were to repeatedly see her visage reflected in a pool of water, there would be no more confusion about self-recognition than when a modern teenager scrutinizes his troubled complexion daily in the bathroom mirror. But what about the other three large primates that also originated in Africa?

In a National Geographic film that always delights my students, a chimpanzee at Jane Goodall's early field station at Gombe National Park, Tanzania, approaches a mirror that has been set up temporarily in its environment and becomes intensely involved with the image it sees. It stares motionless for a moment, and then experimentally the curious ape moves its head sideways several times just to see what its imaged counterpart will do. Suddenly, it whips its head far to one side in order to quickly catch a view of the entity that seems to be positioned behind the mirror. Apparently, it is trying to outwit the "other chimpanzee" that is such a perfect—but elusive—imitator, and it goes through this sequence several times. But each time, of course, the phantom "disappears" from view at the very last minute.

Compare this chimp's naïveté with that of Harpo Marx when he guest-starred in a remarkable episode of *I Love Lucy*. Lucy is dressed like Harpo right down to a curly blonde wig, and for some reason of sitcom logic she is hiding from Harpo in a closet. When he slides open the door, Lucy pretends to be his reflection in a mirror, and Harpo becomes engrossed by what he takes to be his own marvelous image. Soon a merrily narcissistic Marx Brother is clowning up a storm, and Lucy's impersonation is so good that for a long time Harpo is fooled. But eventually she makes a barely noticeable mistake, and a now suspicious Harpo begins to seriously test this "mirrored" image. Some great comedy ensues as Harpo pushes Lucy's imitative powers to the limit—and in the end she is revealed as an imposter.

Such antic hijinks aside, the question remains: Could a properly trained chimpanzee, bonobo, or gorilla be *capable* of recognizing itself in a mirror? The answer seems to be yes, as psychologist Gordon Gallup demonstrated in a series of famous psychological experiments that probed the capacities of various monkeys and apes for self-recognition.[28] First, Gallup allowed his captive subjects to become habituated to mirrors by placing one in their cage for ten days. (This was, of course, an advantage that the curious but utterly perplexed wild ape at Gombe didn't have.) And both monkeys and apes took great interest in their images, but after ten days it was still impossible to tell whether they considered the image to be perhaps a stranger of their own species or themselves.

Next, Gallup applied small dots of red pigment to their faces while they were sedated, placing the dots where they would not be visible without a mirror. (Both monkeys and apes have color vision, and they tend to "react" to the color red on a living being—possibly because it suggests the presence of blood.) Then, he allowed his subjects to wake up completely before he replaced the mirrors in their cages. In response, macaque monkeys at best would show some interest or alarm at the reflection, perhaps as though seeing another monkey that was wounded. But upon first seeing their images in the mirror, often the chimpanzees and bonobos, and eventually to a lesser degree gorillas, immediately and consistently touched the red dot on their own faces as if following a cognitive process along the lines of: That *thing* has to be on *me!* The responsive great apes obviously realized that the image was their own, and subsequent experiments have shown that certain other highly social, large-brained species can do something similar.[29]

Gallup's rather narrow test of "selfhood" confirms what language-laboratory researchers sense intuitively about the apes they work with. They readily learn their own names as well as the names of others, and they are capable of playfully manipulating their images in mirrors, for instance, with makeup. In general, they act as though they understand at some level who they are as social beings, and in fact one female chimpanzee named Washoe, who had been taught to use sign language, seemed to have been convinced that she was a human. Indeed, the first time she saw another chimpanzee she signed, "BLACK BUG!"[30]

In humans it's the individual capacity to understand that there is a *self,* which exists in relation to others, that makes it possible for people to participate in moral communities. Obviously, mere self-recognition does not make for a moral being with a fully developed conscience, but a sense of self is an important and necessary first step. It's useful in gauging the reactions of *others* to a person's behavior and in understanding their intentions. And it's especially important to understand that the person can become the center of a hostile group's attention if his or her actions seriously offend its moral sensibilities. The aforementioned capacity to take on the perspective of others not only underlies the ability of individuals in human communities to modify their behavior and follow the rules being imposed by the group, but it also allows people acting as groups to predict and insightfully cope with the behavior of "deviants."

Even for field researchers like myself who have had the opportunity to observe chimpanzees long term in their natural habitat, it can be difficult to say whether they are "perspective taking"—which means taking into account others' motives and reactions—nearly as well as socially sophisticated humans do. However, in captive experiments a chimpanzee can deliberately engage in perspective taking in order to deceive other apes.

In a truly ingenious experiment, a young male was allowed to see the researcher bury some fruit in a large outdoor enclosure, while the rest of the sizable group remained out of sight. When the entire group was allowed to enter the enclosure, of course the young male went straight for the food.[31] As the experiment was repeated, the older, higher-ranking males soon learned that the young male would know exactly where the food had been hidden, and thereafter they simply watched his movements, and as he began to dig up the food, they—being dominant chimpanzees—chased him away and ate it themselves. As the experiment went on, however, the young male demonstrated that he could get inside the heads of these dominators and outwit them. In later tests, he would rush to a location where he knew there was no fruit and excitedly begin to dig. When the others followed suit, the low-ranking young chimp would unobtrusively move to the actual site of the buried fruit, managing to eat at least some of it himself before his superiors noticed him and rushed over.

Basic elements of perspective taking also are discernible in chimpanzees in situations that are not patently experimental. For instance, primatologist Frans de Waal describes the case of two males in a large captive group who were vying for the dominant alpha position. As they were trying to threaten each other, they both exhibited the characteristic highly visible toothy grimaces that advertise fear.[32] This facial expression is involuntary, but as the contest dragged on, one of the two males cleverly started to place his hand over his mouth to deny his rival the visual cue that he himself was feeling stressed. A somewhat similar behavior has been reported at Gombe, where an excited wild male tried very hard to suppress his own involuntary food calls in order to avoid losing precious provisioned bananas to just-out-of-sight rivals who were alerted by the calls.[33]

We humans resort to deception all the time, and telling a credible lie requires that we fashion our words very carefully to create a well-constructed illusion in another person's mind. To take an extreme instance, a successful bigamist must display these skills to the

n^{th} degree, taking into account many variables such as how suspicious or trusting his (or conceivably her) respective mates are likely to be and what kinds of excuses will fly with each of them. But sly deception is just one of the many uses of perspective taking.

When large-brained animals are evolved to live in hierarchical social groups, they are likely to exhibit a fairly sophisticated basis for assessing the motives of others, either in order to competitively dominate them or simply to survive and try to flourish in a less stressful subordinate role. Perspective taking[34] also can involve reckoning what a group of others may do if an individual's behavior makes her or his peers become so aggressive that they begin to act as an angry, aroused coalition. All of these skills are useful to the individual, and their strong persistence among all four of the African great apes today (technically humans are, whether we like it or not, an African ape[35]) tells us that they have been providing individual fitness benefits for at least 8 million years.

This means that an important head start already existed whenever environmental changes began to stimulate conscience evolution in earlier humans. But taken by itself, this particular preadaptation, the primitive "self," was not in any way moral, for morality involves a pointed sense of right and wrong coupled with a sense of shame and, on the opposite side of the coin, a sense of honor and pride.

FOLLOWING THE RULES—OR ELSE

At a bare minimum, humans and the three other ape species that originated in Africa also know intuitively what a "rule" is. This usually involves a stronger individual's demanding a certain type of behavior or insisting on the absence of an unwanted behavior, and the expectation is backed up by a potentially hurtful authority. However, for a Bushman hunter-gatherer, living in some remote place out on the vast Kalahari Desert, the most salient authority of all will be the egalitarian local group as a whole. The band can be a stern authority, indeed, first because its social expectations are intensified by being moral, and second because such people are in the habit of using weapons to kill mammals the size of humans or larger. This means that an outraged group can become an aggressively hurtful group.

People like Bushmen or Pygmies gossip incessantly and are highly judgmental, and group opinion is something to be feared because moral outrage can lead to ostracism, expulsion from the group, or even execution. This is true of all hunter-gatherers. For instance, as we saw in Chapter 4, if an overbearing individual were to kill one band member and then killed another, as a repeat killer and disliked dominator he'd be dealt with by this collective authority—whether he hunted and gathered out on the Australian semidesert or on an arctic tundra or anywhere else. And he'd be dealt with lethally.

Among modern humans, moral authorities can range from parents at home to the immediate community, represented by a local cop or a county sheriff backing up the law. For a believer, there may also be an omnipotent and punitive God. Such authority, be it "informal," or formalized in statute, or religious, brings with it specific, well-understood rules. We can state such rules in terms as varied as "Thou shalt not steal" or "Trespassers Will Be Prosecuted," or "No TV tonight until you finish your homework." Among hunter-gatherers, the supreme authority is the local group that camps together as a band, and Émile Durkheim has described beautifully the near-tyranny that such groups have over the individuals who are driven to conform.[36]

Among chimpanzees, bonobos, and gorillas, certain basics seem strikingly similar. Dominant individuals easily impose "rules" on their inferiors. For instance, a subordinate knows that if a small prime feeding site is encountered by a foraging party, the rule is *Don't make the first move, or else you'll be aggressively threatened or physically attacked by a proprietary dominant.* Such nonmoral rules have mainly to do with feeding priority or male access to females for mating, but sometimes they arise out of competition for political position, pure and simple. Indeed, in the wild male chimpanzees devote a great deal of energy to seeing who can rise higher in the male dominance hierarchy, and they bear scars to prove it. (Females are far less competitive.)

Compared to chimpanzees, bonobo males seem much less obsessive in competing with other males, but they, too, bear scars, probably mostly from the females who unite to go up against them and contest their authority,[37] while in the wild bonobo females also compete for alpha status.

In all of these contests, the rules are both simple and important to individual fitness. The dominator, acting as an authority, insists on either an immediate tangible concession or a sign of deference or appeasement, and as long as this ensues, things will go smoothly. With humans, often it is "society," rather than a single individual, that serves as the authority, but as we're about to see, the imposition of at least a few "rules" on individuals by sizable *groups* was already under way with the Common Ancestor and even more so with its successor, Ancestral *Pan.*

HIERARCHY CAN BE FLEXIBLE

Richard Wrangham created his portrait of the CA as a group-living killer ape in 1987, and later, in his coauthored book *Demonic Males,* he greatly expanded this social portrait as he moved on to the ape I'm calling Ancestral *Pan* and compared bonobos and chimpanzees in looking for the origins of human violence. In *Hierarchy in the Forest,* when I analyzed the pervasive pecking orders that can be factored into the Common Ancestral equation, it was clear that chimpanzees are so hierarchical that every male knows exactly whom he can back down and whom he must submit to; there's even a specialized subordinate greeting that makes this clear. Among gorillas, huge silverbacks intimidate all the adults in their harem, while the females have their own pecking order. Bonobos' social dominance hierarchies are complicated by the fact that males don't form coalitions and pairs of females regularly gang up against otherwise dominant males and thereby often manage to control the best food. Human hierarchies can be more complicated still, but all of us live in groups with pronounced hierarchical tendencies, and living in such groups requires an understanding of rules.

For humans, there's the special "egalitarian" twist that I emphasized in Chapter 4, and this merits further discussion because it will be crucial to our evolutionary analysis. If we look to the 150 LPA hunter-gatherer bands that most closely resemble more recent prehistoric human societies, we've seen already that they're highly egalitarian. Minimally, this means that all the active hunters (generally the adult males) insist on being seen as equal and that among themselves they tolerate no serious domination—be this in hogging vital food resources or in bossing others around. Based on assumptions coming out of behavioral ecology, I'll soon be making the case that very likely such egalitarianism arose—or was greatly intensified—when our predecessors began to go after large game in a serious way.

We've already seen what happens with the Bushmen, who preemptively put down a self-aggrandizing citizen before he—it always seems to be a he—can get up a head of steam in the direction of building himself up and acting superior. With such people there's still some possibility for upward social mobility in the sense that sometimes a wise individual may be accorded the status of temporary or permanent band leader.[38] However, that person is expected to behave with humility, for the accepted leadership style permits nothing more assertive than carefully listening to everyone else's opinion and then gently helping to implement a consensus—if this spontaneously forms. Such decisions may involve a band's next move or the group's taking action against a serious deviant, but such leaders by themselves cannot settle on an outcome; this is a decision for the entire group.

ANCESTRAL "SOCIAL CONTROL"

Here we have an important topic that I shall merely introduce at this point. Among wild-living great apes, the most pervasive and routinized *group* control of individuals probably is to be seen in bonobos, whose coalition behaviors are interesting because female power is significantly augmented. They live in large quasi-territorial communities and forage in mixed parties that include a sizable number of males and females.[39] The males compete with other males for dominance position by using their mothers as political allies; they never form such alliances with other males as chimpanzees do so predictably. And because bonobo males are both somewhat larger and definitely more muscular than females, any one-on-one contest sees a high-ranking male bonobo as the predictable winner.

However, a female usually has one or more female allies nearby, and two or more of these feisty bonobo females can readily back down a bigger male and take the preferred food. If a male's would-be dominance provokes really strong hostility, up to half a dozen aroused females may exert *serious* control by gang-attacking the offender and biting him severely enough to possibly kill him.[40] This social control by coalitions of female bonobos occurs frequently not only in large wild communities but also in smaller captive groups.

Gorillas may also engage in collective social control in rare instances. The females are only about half the size of silverbacks, and because they live in harems and compete to see who can stay in closest proximity to their huge male protector, these females are not likely to form counterdominant coalitions the way bonobos do. However, captive situations can bring out aspects of a species' social potential that might be seen only very rarely in the wild. Reported is a case in which a young adult blackback male presided over a harem of several females, until one day a huge silverback obtained from another zoo was introduced with the expectation that he would immediately take over the group, dominate it, and adjudicate its quarrels. Instead, the united females attacked him so vigorously that he was left cowering in one corner of the enclosure and had to be removed.[41] In this case the contention was not about food but about who would dominate the group, and this social control by several united females was effective in decisively rejecting the enormous, physically powerful newcomer.

Chimpanzee coalitions exert social control in a variety of ways. In the wild, female chimpanzees, unlike bonobos, have to forage alone much of the time because their food supply is not so plentiful, and for that reason they have inadequate opportunity to form alliances with other females. As a result, unlike the bonobos, the lowest adult male on the totem pole can dominate even the toughest female because it's always "one on one."[42] On the other hand, males are continuously forming small coalitions in hope of unseating the alpha, and this can undermine his power unless he forms effective coalitions of his own. Thus, the males are constantly making use of dyadic alliances to gain a competitive edge.

At times, subordinate male chimpanzees, sometimes with the help of adult females, may decide to collectively buck the disliked authority of one of their community's high-ranking males by ganging up in large numbers to attack him and drive him out of the group. Though rare, this has happened several times at two different field sites in Tanzania,[43] and in one case a rejected bully who had once been alpha disappeared, never to be seen again. As with bonobos, this brings chimpanzees perilously close to practicing capital punishment by group action, and it seems that they make their decisions based on what primatologist Frans de Waal calls "community concern." By this, he means that they *care* about controlling aggressively disruptive individual behaviors that potentially harm the interests of other community members.[44]

When chimpanzees live in large captive groups, things most definitely change for the females. Being well provisioned, they no longer need to go out alone to forage, which means they can now form strong political bonds with other females—just as female bonobos are able to do in the wild because their food is plentiful and their foraging parties are so large. These formidable female captive chimps go beyond their bonobo counterparts as they join together in still larger counterdominant coalitions that make serious inroads into male power. They do so by enforcing a few simple rules. These united females are most prone to gang-attack a male who wants to take out his frustrations by beating up the nearest female—a behavior that is quite routine in the wild. Captive females not only control such male aggression against females, but, by reacting as a whole community, they also often can control male bullying in general.

Frans de Waal describes such an incident, which involves the alpha male's becoming aggressive against a much lesser male:

> Jimoh, the current alpha male of the Yerkes Field Station group, once detected a secret mating between Socko, an adolescent male, and one of Jimoh's favorite females. Socko and the female had wisely disappeared from view, but Jimoh had gone looking for them. Normally, the old male would merely chase off the culprit, but for some reason—perhaps because the female had repeatedly refused to mate with Jimoh himself that day—he this time went full speed after Socko and did not give up. He chased him all around the enclosure—Socko screaming and defecating in fear, Jimoh intent on catching him.
>
> Before he could accomplish his aim, several females close to the scene began to "woaow" bark. This indignant sound is used in protest against aggressors and intruders. At first the callers looked around to see how the rest of the group was reacting; but when others joined in, particularly the top-ranking female, the intensity of their calls quickly increased until literally everyone's voice was part of a deafening chorus. The scattered beginning almost gave the impression that the group was taking a vote. Once the protest had swelled to a chorus, Jimoh broke off his attack with a nervous grin on his face: he got the message. Had he failed to respond, there would no doubt have been concerted female action to end the disturbance.[45]

As with a small human band that has its hackles up and is bent on social control when an important social rule has been broken, the chimpanzees' escalating hostile vocalizations may be seen as public opinion heating up as the group watches to see if a serious deviant is going to change his behavior. The female chimpanzees cease only when the alpha male stops the unwanted behavior, and this tells us something else: that great apes are not intent on totally eliminating dominant individuals whose behavior they strongly disapprove of.

They are willing to give them a chance to cease and desist—in human terms, to reform. Later, I'll be producing some statistics on LPA human social control, which show that we, too, prefer to reform deviants rather than kill them.

What, then, was the CA's capacity for collective social control? We must go to the least common denominator, which is the gorilla. We can conclude that this ancestor's hierarchical behaviors included at least the *potential* to gang up aggressively and inflict wounds, with aroused coalitions coalescing to control the behavior of powerful individuals whose actions were a source of serious and *shared* irritation and hostility. And if we look to Ancestral *Pan,* then the least common denominator becomes bonobos and chimpanzees, whose counterdominant behaviors are much more frequent than with gorillas. As we'll be seeing in the next chapter, this potential for rebellion was very important, for it was new developments in the practice of group social control against dominant individuals that set humans on the course to evolve a conscience.

THE EVOLUTIONARY CONSCIENCE

To summarize, we now have before us a group-living, decidedly hierarchical Ancestral *Pan,* an ape that had a sizable prefrontal cortex and at least a fairly complex social self. It had a matricentric family that permitted the young to learn from the old, and part of what was learned were rules of behavior—rules that pertained to social competition and, politically, to the dominance, submission, and coalition-formation moves that such competition engendered. This ape had quite a sophisticated capacity to understand the manipulative intentions of others, whether they were acting as forceful, high-ranking individuals or as hostile, dominant groups, and in certain restricted situations it obviously was capable of exerting some social control on a collective basis. But do these capacities add up to anything like what we might call a conscience, with a moralistic sense of right and wrong?

Having a conscience is all about *personally identifying with community values,* which means internalizing your group's rules. You must not only be able to learn rules and predict the reactions of those who enforce them, but you also must *connect* with these rules emotionally. You must do this in a positive way that makes you identify with them, feel ashamed when you break them, and feel self-satisfied and moralistically proud when you live up to them. This last can be considered a modern definition of virtue.

For a number of reasons, individuals who better *internalize* their groups' rules are more likely to succeed socially in life and thus be more successful in propagating their genes. On the opposite side of the moral-conformist coin, serious inabilities to identify emotionally with group rules are likely to reduce personal fitness, as is the case for the many sociopaths who are liable to be incarcerated today, unless they are unusually adept at avoiding detection, and who in the past would have quickly run afoul of their vigilant band's moral system since all band members act as moral detectives. For humans, fitting in with your moral community has a high fitness payoff because being punished is costly to fitness, whereas having a good reputation can help fitness.

By definition, moral communities are groups whose members have a sense of right and wrong based on rules being internalized, and the result is collective preferences that affect people's reputations and hence their moral standing. When groups of individuals have internalized the same set of rules, they can implement passionately judgmental social control against serious deviants if all agree that the antisocial actions in question are shameful or monstrous or very threatening. It's the conscience, as a moral compass, that orients personal behavior in the context of this group life, which involves both punishment and rewards.

I've already suggested that group rules should not be internalized so strongly that you'd be free of any temptation to break them, for many of the prohibitions that human groups arrive at are designed to curtail the same selfish behaviors that—in smaller doses—can help individuals to advance their reproductive success. What's biologically optimal, I've suggested, is to identify with rules to the degree that, when you're about to *seriously* transgress, you'll experience a sharp feeling of psychological malaise and possibly a burning in your cheeks. This reminds you that a major social covenant would be broken and that there could be serious social consequences.

Thus, as socially well-adapted people, we are not totally ruled by our consciences. Far from it. Rather, we are informed by them, and we are effectively inhibited but in a *flexible* way. That's how most of us manage to make some minor moral compromises in order to get ahead in a competitive world, yet basically maintain a decent reputation and stay out of really serious social trouble.

DO APES HAVE MORALS?

As our conscience helps us in sorting out life's moral dilemmas, a fair portion of its operation can become conscious. For instance, many of us can actually hear ourselves talking to ourselves as we face a moral dilemma and weigh the consequences, as Darwin apparently did. Is anything even remotely similar to this Darwinian "inner voice" making itself known in the mind of an ape? This question isn't as outlandish as it may seem, for we'll be seeing soon that captive apes, when alone, do at least talk to themselves in sign language about certain kinds of things.

In leading up to this question, we've already reconstructed a "rule-oriented" ancestral capacity to exert punitive social control and to respond efficiently to such control in terms of fear and also a fairly sophisticated ancestral sense of self-awareness. The next question is, were our distant ancestors inhibited by anything like *"moralized"* feelings of shameful inappropriateness, based on an internalization of values and rules that results in an inner sense of right and wrong? If chimpanzees, gorillas, and bonobos were capable of such feelings today, even at a rudimentary level, and if they were capable of internalizing rules as we do, with self-inhibition resulting, then we would have to conclude that our Common Ancestor was already a moral creature of sorts. In that case, in all fairness, humans would not be the only moral beings in the world—and theories of moral origins would have to reach back over millions of years.

How can this hypothesis be tested? If as an experiment a few gorillas, chimpanzees, and bonobos were raised by moralistic humans, and if as a result all three species unanimously showed some definitive evidence of internalizing rules and harboring a sense of shame as we do, then we could say that in its potential our Common Ancestor was already well along the way to evolve a modern conscience. If they showed signs of being *privately* concerned about their own behavior when they were breaking rules, and nobody else knew about it, this evolutionary head start would be stronger still. To explore these important questions, our journey in search of a scientific Eden will now take us into a most unusual type of laboratory.

CAN AN APE BE "GUILT-TRIPPED"?

Since 1959, a small number of captive chimpanzees, along with a few bonobos and gorillas, have been raised by humans and trained to use American Sign Language (ASL) or, more rarely, to become fluent in using other manual communication systems devised by humans. Mainly, these highly sociable animals are taught to use symbols that will manipulate their masters, as in asking for food, but because discipline is needed to maintain order in the lab, a few items in their limited vocabularies are designed to help in keeping them under control.[46] When apes are taught to use social interaction signs such as GOOD and BAD or SORRY, at least in the minds of their masters these signs designate morally loaded ideas and feelings.

The apes in question are not merely "domesticated" in the sense of being dependent captives who are denied a natural freedom of movement. Many of them are also cross-fostered, which means that basically they have been raised by humans much as people raise their own children. This is feasible because small apes are innately very similar to human children, being intelligent, curious, affectionate, sociable, and, initially, extremely dependent. They are also compulsively playful, and it is in their nature to be communicative.

The majority of these laboratory animals have been chimpanzees, and their scientist trainers have to take a special care in this unusual enterprise because their subjects are equipped with serious and mercurial tempers. They also have powerful jaws, with large canine teeth to match. Furthermore, even as relative youngsters they will have become several times as strong as humans. Fortunately, their volatile aggressiveness and the accompanying tendency to willfully make mischief are combined with a disposition to submit to domination. But with their physical power and aggressiveness, their social tractability can never be taken for granted.

For obvious reasons, in both homes and laboratories these cross-fostered apes experience toilet training, which involves either positive or negative reinforcement. They also experience praise or blame with respect to other behaviors that please or displease their moralistic masters. Here's the question: Does this special, humanized kind of socialization have any unusual effects in bringing out a captive ape's behavioral potential—effects that might provide some hints about a much earlier, ancestral potential for moral behavior?

We should begin by first asking how ape mothers control their impetuous offspring under natural circumstances. Year after year, at Gombe National Park, I watched carefully to see how wild chimpanzee mothers socialized their infants, and, as this was Jane Goodall's special research focus, I was privileged to spend hundreds of hours observing mother-infant interactions while I, a cultural anthropologist by training, was learning how to be an ethologist. As Jane has written,[47] a good chimpanzee mother seems patient almost to the point of martyrdom, and basically she responds to her offspring's cues in ways that appear to be hostility-free or, in human terms, "nonjudgmental." The mother protects and controls, but she does not appear to angrily condemn or blame her infant. Tolerant guidance is the name of her game.

For instance, when the infant attempts to nurse, the mother's nipple is readily made available. As the infant grows larger, its begging gestures may divert food from its mother's mouth. If two youngsters are roughhousing and one begins to inflict some pain, the mother tends to intervene impartially, protecting the victim but punishing neither party. Perhaps once in a long while the mother's ire may appear to rise a bit if a youngster becomes pesky, but basically the manner in which she treats her offspring is calmly protective and quite far from being hostile or "disciplinary." Of course, if another adult chimpanzee or some predator threatens her precious charge, at risk to herself she will defend it forcefully and with real hostility. But when she controls her offspring, anything reminiscent of moralistic judgment—or even simple anger or resentment—seems to be absent.

In our own species, moral socialization by parents can involve severe disapproval and shaming and sometimes physical punishment or possibly restriction of freedom. We'll see this in later chapters when I quote at length from the autobiography of a voluble female member of a Kalahari foraging band. With chimpanzees it seems to be simply a matter of firm, friendly, dominant guidance. And after six years of annual visits to Africa, my conclusion had to be that no parental behavior existed that seemed likely to stimulate a "moral self" in an immature wild chimpanzee. Inspired by Darwin, I was looking for any sign of an incipient ape conscience—and I didn't see it. I was left, however, with the possibility that perhaps such a behavioral potential does exist, but under natural wild conditions its development simply is not stimulated.

To explore this, we must turn to these captive subjects, whose human "parents" just naturally are morally judgmental. Apes make creative use of the limited "languages" we devise for them, and they use them to hold meaningful, two-way conversations with humans and even with other apes. Indeed, presently I shall relate how a willful gorilla and I got into an acerbic argument that is quite easy to follow, and I was insulted in a way I will never forget.

Great apes readily learn ASL,[48] while another system, used for decades in a laboratory in Atlanta with chimpanzees and later with bonobos, has been facilitated by a computer console that has totally arbitrary symbols on the keys.[49] In both of these "manual" systems, the potential vocabularies are quite similar in size and content. The lexicons in question are curtailed by the cognitive limitations of African great apes and also by the restricted life circumstances of apes in captivity, yet they usually include at least one hundred different "signs." Younger apes make the best experimental subjects, for they have predictable social and physical needs just as human children do. The vocabularies that psychologists devise for them take this into account, and a fair number of the signs or symbols signify things that little apes enjoy enormously, for instance, different types of food and drink or attractive concepts like play, and specific activities such as hug, or tickle, or chase. Because these animals have self-concepts, they can also be taught their own names as well as the names of significant others, ape or human.

There are also signs for asking questions, for these animals are just naturally curious. For instance, in the 1970s psychologist Maurice Temerlin and his wife raised the chimpanzee Lucy from infancy with their own son, treating her as much as possible like a human child. Temerlin writes of Lucy, "As she sees something new she often asks 'WHAT'S THAT?' by moving a forefinger rapidly left and right (WHAT) and then pointing the same forefinger at the object to be identified (THAT). She asks this question of us, and at times of herself, as she leafs through a magazine and sees something she has not seen before. We are sure she is talking to herself or asking rhetorical questions when Jane or I are too far away for her to be talking to us, or even out of sight."[50]

Unfortunately, however, none of this ape communication is powerful enough to allow us to conduct ethnographic interviews—to directly query apes about whatever "moral" feelings they might have. We must rely instead on inference, and there is one telling gap in the ape repertoire that is wholly unambiguous. Apes don't blush for reasons that are social, whereas as Darwin demonstrated, humans everywhere do this. We blush with embarrassment, and we also blush with shame. No other species does it; even a cross-fostered ape like Lucy lacks this response.

Let's say a chimpanzee infant is being raised by people; no matter what their disciplinary strategy may be, they'll naturally tend to treat it as they would a human child, giving signs of approval or disapproval as they toilet train it and otherwise regulate its exuberant behavior. The humans' moralistic cues may be obvious or subtle, but they'll be there. As it masters its vocabulary, eventually the small ape will have been exposed to judgmental signs like GOOD, BAD, and SORRY with respect to its own behavior. In an important sense, we can say that it is being socialized to be a moral being—so it's of interest to see how much potential the chimpanzee may have to respond in kind.

Unfortunately, the vocabularies used with chimpanzees are not designed to differentiate between just plain GOOD, as in good tasting or fun to play with, and GOOD in the sense of something that is *morally* appropriate, as in "It's good for you to be generous." Thus, it's hard to imagine how an ape will interpret the command "GO POTTY GOOD." The same goes for moral BAD, as in a trainer's signing that defecating on the floor is BAD. How do we know whether the ape is thinking in terms of "*practical* BAD," meaning you get scolded or punished if you do it, as opposed to "*moral* BAD," meaning that intrinsically the act is antisocial and therefore the actor is doing something naughty or shameful? The distinction is rather subtle, but very important.

The same goes for SORRY. For humans, "sorry" can indicate deeply felt remorse tinged with self-blame and shame or merely simple, nonmoral regret.[51] The problem, again, is that these talented apes are given only one SORRY sign to work with. With respect to SORRY, during my research in the African forest the apes never appeared to me as though they were upset over their own behavior, let alone ashamed of it or remorseful. I did notice that between individuals there were postures and gestures that seemed to ask for or grant forgiveness, and in fact chimpanzees often make up after conflicts.[52] Basically, one ape simply touches the other's hand or body—sometimes unilaterally, sometimes mutually, and sometimes, as observed in captivity by de Waal,[53] with the useful assistance of a third party. However, this seems to be aimed merely at reducing tension or restoring positive relations, so reading a morally based element of remorse into such behavior would be patently *anthropocentric.* Nothing I observed ever convinced me that there was something like morally based self-recrimination in the wild, for aggressors never appeared to be troubled by their actions afterward. But what about these captive apes?

When Lucy was nine, she had worked with sign language her whole juvenile life and firmly controlled one hundred signs. At this time, the single sign that was really suggestive of moral connotation was SORRY, for GOOD and BAD hadn't been added to her vocabulary at that time. There was also DIRTY, which presumably had played a major role in her toilet training. During this period, psychologist Roger Fouts was visiting to tutor Lucy in signing, and shortly before a language lesson, when no one else was in the room, Lucy had defecated right in the middle of the Temerlins' living room floor. When Fouts noticed the "crime," he turned to Lucy, and here's their conversation in ASL, "verbatim." Fouts begins with a moralistic, accusatory confrontation.

> *RF:* What is that?
> *Lucy:* Lucy not know.
> *RF:* You do know. What's that?
> *Lucy:* Dirty, dirty.
> *RF:* Whose dirty, dirty?
> *Lucy:* Sue's *(a graduate student, not present).*
> *RF:* It's not Sue's. Whose is it?
> *Lucy:* Roger's!
> *RF:* No! It's not Roger's. Whose is it?
> *Lucy:* Lucy dirty, dirty. Sorry Lucy.[54]

Lucy's efforts at "tactical deception" in placing responsibility for the mess elsewhere might be seen in the same category as pretending to find the buried fruit or covering up a fear grin,[55] and though her attempt was clumsy, the intention was obvious enough. In theory Lucy's initial try at blaming someone else *might* have worked because Sue wasn't present. Her second try, in its ridiculousness, suggests that a stressed young chimpanzee was getting confused and running out of likely scapegoats.

Temerlin recounts other instances of Lucy's attempts to deceive her masters, as when she "fails to understand" a sign that she knows extremely well. Indeed, LUCY NOT KNOW was the first line of defense when Roger came after her. It's clear that Lucy did understand what was going on in that she realized that she was being held accountable for a past misdeed. Sensing that she had broken the rules, with her lies and subsequent "apology" apparently Lucy was trying to get herself off the hook with a major authority figure. But was there some unobservable element of negative self-judgment or shame in this symboled apology? I very much doubt it.

If we compare Lucy's nonmoral response to the equally nonmoral reactions of highly trainable domesticated dogs, a similar capacity to learn rules from humans is evident. But Lucy's reactions betray an understanding that a past action goes against the rules, which involves a capacity to understand ex post facto punishment that dogs don't seem to possess. Of course, when disapproving human groups sanction deviants, they often do so *long* after the fact, and to good effect, for the deviants understand exactly what is going on. Likewise, chimpanzee Lucy not only understood her past culpability but also was able to connect it with future punishment. She was even sophisticated enough to try for a "cover-up," or so it would seem.

The same may be true of bonobos. As a young bonobo male, Kanzi was taught to use a computer keyboard (instead of sign language) by psychologist Sue Savage-Rumbaugh in her laboratory in Atlanta, where I had the opportunity to observe him over several days. Singlehandedly, Kanzi has shown that bonobos have at least as much linguistic capacity as chimpanzees, and with respect to the questions of "morals," Kanzi's trainer writes:

> When the lexigrams GOOD and BAD were first placed on Kanzi's keyboard, I did not think he would use them frequently, or with intent. I put them on so everyone would have a clear way of indicating to Kanzi when we felt that he was being good or bad. To my surprise, Kanzi was intrigued with these lexigrams and soon began using them to indicate his intent to be GOOD or BAD, as well as to comment on his previous actions as GOOD or BAD. When he was about to do things that he knew we did not want him to do, he started saying BAD, BAD, BAD before he did them, as though threatening to do something he was not supposed to do. He would, for example, announce his intent to be bad before biting a hole in his ball, tearing up the telephone, or taking an object away from someone.[56]

Sue also relates how Kanzi labeled his treatment of another researcher:

> One day, when Kanzi was supposed to be taking a nap with Liz, who was exhausted and went to sleep, Kanzi refused to lie down. After she had been asleep about fifteen minutes, she suddenly realized that the blanket she was using as a pillow had been rudely jerked out from under her head. She sat up to look over at Kanzi who commented on his action, saying BAD SURPRISE. Another time, when he was supposed to take a nap, he did not want to do so. He asked to play CHASE WATER instead, and when told he could not do so, he commented BAD WATER and proceeded to take the water hose and spray it all over things.[57]

It's apparent that this bonobo was able to place labels humans will take as moralistic on his own behavior, with some degree of understanding, and he did so not only with respect to past behavior but also with respect to future behavior. However, using human labels does not necessarily mean that he experienced our own type of moral feelings. Indeed, there is no hint that Kanzi was feeling remorse or shame when he announced the BAD SURPRISE or when he signaled that he was going to do something BAD and then went ahead and did it. The difference between knowing you've previously misbehaved in someone else's eyes, which may involve some anticipated sanction or punishment, and feeling inside that you've behaved *immorally* and deserve to be punished is profound.

Lucy was literally a second child in the Temerlin household. Washoe, the first chimpanzee to be taught ASL, was raised less intimately by humans. Comparing Washoe and Lucy, Fouts writes:

> Whenever Washoe's antics tested my patience, I used to conjure up an imaginary BLACK DOG to scare her into cooperating. Maury Temerlin, ever the psychotherapist, would manipulate Lucy with guilt. This was remarkably effective. If Lucy was refusing to eat her dinner, Maury would plead, "For God's sake, Lucy, think of the starving chimps in Africa." She'd then take just a bite or two. Unsatisfied, he'd beg, "Take at least three more bites for your poor suffering father who loves you." Lucy would eat with a little more enthusiasm. Finally, when Maury whined, "Lucy, how could you do this to me?" she became putty in his hands. After a few years, Lucy developed a guilty expression that immediately gave her away whenever she was hiding a key, smuggling a cigarette lighter, or committing some other household crime.[58]

We must ask immediately whether Lucy had the slightest idea what her clinical psychologist father figure meant when he referred to her own species starving in Africa. Cross-fostered chimpanzees do pick up some spoken English in the sense of passive understanding, but such complex concepts would not be represented in that vocabulary. Surely what was affecting Lucy, as an empathetic, rule-responsive being like ourselves, was simply Temerlin's tone of voice.

Temerlin was, in fact, actively trying to make Lucy feel guilty. And chimpanzees, wild or captive, are very sensitive socially. They certainly understand the difference between another's positive or negative feeling toward them. More specifically, they understand the difference between a significant and dominant other's approval or disapproval of how they are behaving or have behaved recently. Lucy probably understood that Temerlin disapproved of her not eating, but did his tone of voice actually evoke something like a pang of conscience—in the form of self-recrimination?

In his book about Lucy, psychotherapist Temerlin reported that this "humanized" chimpanzee would appear furtive when she was breaking rules even if she were unaware of the observer, with the implication that guilt feelings might be present. However, a conservative scientific interpretation would be that all Lucy was doing was reacting *politically* to an awareness that powerful people she was bonded to would disapprove of her behavior and be angry or punish her if she got caught. At Gombe, I saw simple furtiveness all the time. For instance, an estrous female would sneak into a ravine to meet an adolescent male there and copulate with him when the alpha male was feeding nearby. They both knew that the alpha was likely to threaten or attack any lower-ranking rival if he saw him consorting with a female that really interested him, so they were taking their brief pleasure (averaging all of eight short seconds) safely out of sight.

In this context the furtive looks were eloquently fearful—but I'd be willing to bet that they weren't "guilty" in the sense of moralistic self-censure over having broken internalized rules. Indeed, if the same male and female were alone when they met, the furtive body language would be totally absent. My belief is that Lucy's "guilty looks" were similarly motivated; they came from a fear of being discovered in the face of predictable negative reactions. So when Temerlin tried to guilt-trip Lucy into eating her food, by complying she was merely trying to accommodate the feelings of a closely bonded significant other. Empathy, most likely, but guilt, no.

As for gorillas, psychologist Penny Patterson's very imaginative work with the well-known female Koko and Koko's male counterpart, Michael, suggests that these apes have a similar capacity for self-conceptualization, similar signing capacities, and a comparable ability to use signs for GOOD and BAD in labeling their own behaviors.[59] After I began my work with Jane Goodall, I was taken one day by a mutual friend to meet Penny and Koko, and because Koko liked to play "cover-up" games, we brought along an old sheet as a treat for this huge young female. Penny set me up face to face with Koko and quickly taught me two signs. One was PICK UP, and the other was BLANKET.

As I gazed into enormous gorilla eyes, I moved my hands to ask Koko to pick up the sheet. I could see that Koko was watching my hands and that afterward she seemed to be thinking for a time. Then she surprised me. She placed a sizable gorilla forefinger in her enormous, wide-open mouth, rested it on one of her teeth, and held it there while looking more or less in my direction. Being an ASL ignoramus, I was bewildered. Penny explained that apparently Koko wasn't in the mood to play cover-up games, but she was fascinated by people's fillings and was signing the equivalent of SHOW TOOTH.

I must confess to more than a normal share of shiny amalgam fillings, and I obliged her request. Considerately, I tilted my head so that Koko could see the lower left quadrant, where the ravages of youthful sugar consumption were particularly prominent. Koko looked and looked, and finally I decided to close my mouth and get back to business. Our next exchange went like this:

CB: PICK UP BLANKET.
Koko: SHOW TOOTH.
CB: PICK UP BLANKET.
Koko: SHOW TOOTH.

Again I conceded. This time it was the upper left quadrant, which fascinated Koko just as much as the lower left even though it was a bit healthier. Again, after what seemed to me to be a very long time, I closed off what obviously was a magnificent vista and went back to signing.

CB: PICK UP BLANKET.
Koko: SHOW TOOTH.
CB: PICK UP BLANKET.
Koko: SHOW TOOTH.

This time we did the lower right, which had only a single filling. But unfortunately this modest exhibit didn't dampen Koko's interest, so again it was I who terminated the inspection. And again, I made my request:

CB: PICK UP BLANKET.
Koko: SHOW TOOTH.

This time I caved in immediately, and unfortunately the upper right quadrant proved to be at least as fascinating as the others. I submitted to pongid scrutiny for an even longer period, hoping that Koko would get tired of teeth, but her interest showed no sign of flagging. Again I closed my jaws, but this time I was determined to stand my ground.

CB: PICK UP BLANKET.
Koko: SHOW TOOTH.
CB: PICK UP BLANKET.
Koko: SHOW TOOTH.
CB: PICK UP BLANKET.
Koko: SHOW TOOTH.
CB: PICK UP BLANKET.
Koko: ————!

Koko had made a much more rapid sign that I didn't understand, and at this point Penny Patterson burst out laughing so hard that it was moments before she could tell me that Koko had just called me a—TOILET!

Koko and I had not only had a conversation; we'd also had a running, back-and-forth *argument*, and in the end, when I firmly stood up for my point of view, I was rewarded with an insult. At least, that's how *I* felt. Maybe I deserved this nasty epithet for giving a poor, dentally curious, signing gorilla a hard time, but here's the important question: In her use of the symbol TOILET, was Koko merely expressing anger and disapproval at my behavior? Or was there a *moral* element to her name-calling that made me and my antisocial behavior *shamefully* bad?

To figure this out for sure may require a brilliant experiment such as those devised by Gordon Gallup to probe self-recognition. But I doubt that Koko's intentions went beyond angrily associating me with a toilet, an object that her interactions with hygiene-conscious humans surely had told her had strong negative connotations. In this connection, Penny Patterson says, "Koko's basic nature is fastidious. She has always hated stepping in dirt: outdoors she will insist that she be carried over puddles—if she can find someone to carry her—and indoors she will scrub and clean her quarters with a vigor that suggest more than mere imitation. Interestingly, the word *dirty,* which she first used at about age three, and which we use to refer to her feces, became one of Koko's favorite insults. Under extreme provocation she will combine *dirty* with *toilet* to make her meaning inescapable."[60] Apparently, my nasty behavior was an extreme provocation, for it qualified me to receive one of Koko's worst epithets. In human terms, she'd at least called me a "shithead."

Based on all these examples, two things can be said about the fascinating communication behaviors we've seen here. First, chimpanzees and bonobos can take symbols that for us have moral connotations and regularly use them in ways that suggest they do understand *something* about rules and rule breaking when they are dealing with us as authority figures. However, for a scientist it's very difficult to read a humanlike internalization of moral values, or more generally anything like a conscience and a sense of shame, into the behaviors of these apes. In fact, Kanzi seems to have *cognitively* understood his rules very well—without identifying with them emotionally.

This suggests that the responses of apes may be something like those of human psychopaths in that psychopaths understand rules but cannot identify with them—they don't emotionally bond, as it were, with the group standards they grew up with. Another similarity is that both apes and psychopaths are prone to dominate and control. What makes the apes *unlike* psychopaths, however, is that the apes are capable of sympathetic feelings for others, based on understanding how emotions work. Full-blown psychopaths aren't, and unlike the rest of us, they're simply born that way. Apes too are born nonmoral, it appears, but in my opinion they are definitely capable of understanding another's internal state by emotionally identifying with it.

SUMMING UP: ANCESTRAL HEAD STARTS

Overall, in home-raised captive apes the evidence for anything very much like moral feelings as we know them is tenuous at best. If we combine what has been discussed here with what can be observed in the wild, in the absence of ingenious experiments I think we may assume that the CA did not have a moralistic sense of right and wrong as we know it today. This ape did gang up to exert social control, but my assumption is that it did so because a bully's actions could stir simple but strong resentment and active hostility among subordinates. That is quite different from moral indignation that arises when an antisocial behavior, considered to be *shamefully* deviant by local mores, arouses a human group. And when premoral ancestral individuals responded to such group hostility, it was fear—and not internalized rules combined with an individual sense of shame—that made them responsive.

Yet as Jessica Flack and Frans de Waal have argued with respect to empathy and perspective taking,[61] it's also clear that many other building blocks useful to the evolution of a conscience were firmly in place ancestrally, and that they persisted over evolutionary time because they served the reproductive interests of individuals. These advantages ranged from the capacity to differentiate oneself from others, to the perspective-taking potential that permitted ancestral apes to get into the heads of other apes sufficiently to understand, in many contexts, what they were feeling and intending.

This highly social type of psyche made it possible to both follow and impose rules, and if angry, when rebellious subordinates lacked the power to impose rules on their own, they could still join in with others, and wield dominant collective power. At the very least, the result was a primitive form of group social control, which had significant fitness consequences because it could result in wounding, expulsion from the group, and, surely, the occasional death. Such antihierarchical behavior comes very naturally to chimpanzees and bonobos as well as to humans, and basically it can be attributed to resentment of being subordinated, pure and simple.

These various head starts suggest that Ancestral *Pan* had plenty of preadaptive wherewithal for evolving a conscience if the right environmental changes came along, but this doesn't mean that evolving a conscience was inevitable—or even very likely. As I've pointed out, the bonobo and chimpanzee lineages have shared this ancestral behavioral potential continuously over the past 6 million years and yet, despite their keen awareness of social rules and the existence of united attacks by angry subordinate coalitions, and in spite of their ability to recognize themselves in mirrors and understand the intentions of others, not one of them has ever been seen to self-judgmentally blush with shame.

How did humans come to do this so predictably all over the world? Much of the answer has to lie in biology. If we are to explain conscience evolution, I believe that initially we must look to natural selection processes that came to favor individuals who had the advantage over their fellows in the matter of controlling their own aggressions, for that is one very basic job that an evolutionary conscience does. If an individual lives in a society where subordinates can gang up and hurt the fitness of dominant individuals who rub them the wrong way, obviously such self-control can be quite useful to fitness.

After thinking for years about environmental changes that might have triggered such selection processes, I have decided that it could have been not so much a change in the physical environment itself as a change in what human groups did to exploit it—specifically, large-game hunting—that helped to do the trick. As we'll see, this particular ecological pursuit would have made some very specific demands on the people involved in terms of major social adjustments that would allow such a subsistence practice to flourish.

In considering the actual selection processes that might have supported the emergence of a conscience, I shall go beyond Darwin's rather tentative byproduct theory, and even beyond much of contemporary evolutionary thinking, to create a distinctive and multifaceted version of "social selection" theory as an explanation for the rather unusual set of agencies that created this moral faculty for us. In the next two chapters, my hypothesis will be that if earlier humans hadn't already had an ancestral head start in the form of nonmoral group social control, today we might be just as amoral as any of the great apes that still live in sub-Saharan Africa, species that continue to live a life based on dominant power and still do so in top-heavy hierarchies that, compared to ourselves, are only rather moderately modified by subordinate rebellions.

Our human ancestors also lived in Africa, and for that reason our search for a scientific Eden will take us not into a lush paradisiacal garden perhaps at the confluence of the Tigris and Euphrates rivers, with trees bearing tempting fruit on all sides, but rather out onto the hot dry plains of Africa where wild game abounded. There are very good reasons to believe that these large mammals had to be killed and carefully shared by entire groups of people—and not just by individuals or small "families"—if our ancestors were to depend on their meat. And this, I believe, will provide an important key to explaining the evolution of morals.

A Natural Garden of Eden 6

A REALLY SERIOUS HUNTER ARRIVES

Our moral origins story begins 8 million years ago, plus or minus, when the lineage of the Common-Ancestor figure that Richard Wrangham identified for us was, as we've seen in Figure 1, splitting in two. At that branching point one of the two new lineages led to today's gorillas, and to judge from what these apes do in the wild, this gorilla-based lineage either developed or retained a basically vegetarian approach to eating, along with a lack of territorial defense of natural resources and a haremic social structure. This will not concern us here, for it's our more recent direct predecessor, Ancestral *Pan,* that we're interested in because of its more developed talent for social control.

It's very likely that Ancestral *Pan* continued all of the CA behaviors we've met with. To know with a high degree of probability what else this more recent ancestor was doing behaviorally, all that's needed is to look for major patterns of behavior that are present unanimously in bonobos, chimpanzees, and humans—but not in gorillas. Again, the evolutionary principle of parsimony applies in reconstructing the behaviors of this ancestor,[1] and, again, it will be useful to keep the analysis as conservative as possible since we are working with a small clade of only three species here. For that reason I will be looking for the least common denominator, wherever unanimity exists, and projecting only it into the past.

TERRITORIALITY, XENOPHOBIA, AND MORALITY

Thus, it will be Ancestral *Pan* who is the original inhabitant of our Natural Garden, while eventually we will have to include archaic *Homo sapiens* as well. For Ancestral *Pan,* the existence of natural-resource defense was significant.[2] This became humanly important because territoriality and warfare pose such a profound practical problem for our own species.[3] From the standpoint of evolutionary biology, such conflict between groups might also have contributed to moral evolution–if the group selection Darwin spoke of was activated sufficiently to have significant effects.[4] But to start with, we must ask exactly how "territorial" and xenophobic the three extant descendants of Ancestral *Pan* are, respectively, and then look for the least common denominator.

Wild chimpanzee communities predictably stalk and kill their neighbors, and they may eventually wipe out an entire group and take over its resources.[5] LPA human foragers, although often fairly peaceful, sometimes have intensive warfare that also rises to the level of genocide.[6] However, bonobos' territoriality is a far paler version of what chimpanzees and humans do, even though a similar basic pattern is readily discernible.[7] When sizable foraging parties from neighboring communities meet near the edges of their territories, the xenophobic males are prone to vocalize hostilely with the smaller party withdrawing, and in one instance an injured bonobo male was observed afterward,[8] so apparently the bonobos' intergroup antagonism is not limited entirely to bluffing. On the other hand, some bonobo groups are reported to join up together amicably.[9]

With bonobos as the least common denominator, this means that Ancestral *Pan*'s territorial tendencies were setting up groups to actively dislike their socially distant neighbors, with at least some possibility of limited physical conflict. Conservatively, with this least common denominator Ancestral *Pan*, though xenophobic, was not a serious warrior.

It seems likely that contemporary very inconsistent LPA forager warfare levels may not be representative for frequent junctures in the Late Pleistocene, when our predecessors were facing dire scarcity due to climate change and group competition was likely to have been seriously exacerbated.[10] One of the remarkable things about xenophobic tendencies in contemporary hunter-gatherers, and for that matter in all humans, is that our moral codes apply fully only within the group, be it a language group, a nonliterate population that shares the same piece of real estate or the same ethnic identity, or a nation.[11] There seems to be a special, pejorative moral "discount" applied to cultural strangers—who often are not even considered to be fully human and therefore may be killed with little compunction.[12]

This moral downgrading of out-groupers generalizes beyond warfare between groups of coresident armed males, for today it can feed not only into genocide against the helpless, but also into terrorism that targets civilians. More generally, standard military ideologies make the "necessary" killing of civilian enemies palatable, even if they aren't being used as human shields. "Regrettable, but acceptable" seems to define the collateral damage situation unless inflicting civilian casualties becomes a deliberate instrument of national policy, as with indiscriminant World War II bombing of cities like Nanking, London, Dresden, and Tokyo, with enormous and deliberate civilian losses, or as with the American attacks on Hiroshima and Nagasaki, which may have destroyed military targets but also killed similar numbers of civilians.

The underlying raw xenophobia can be traced directly back to Ancestral *Pan*. However, once culturally modern humans began to "moralize" their fear of or contempt for strangers, this gave us ethnocentrism.[13] This culturally refined motivating force helps to support intensive conventional warfare and conquest, along with genocide as a particularly destructive type of warfare. Ethnocentrism, with its moral condescension factor, obviously has been important in establishing the sometimes quite deadly warfare patterns that some hunter-gatherers engage in today, but it may be difficult to estimate exactly how far back this violent, morally based type of cultural behavior reaches, how widespread it might have been in the Late Pleistocene at times when climates were more favorable,[14] and, at the bottom line, to what degree it could have boosted Pleistocene group selection that favored the evolution of altruism.[15]

I must emphasize that the Pleistocene was different, even though conditions that would have spurred warfare were not constant. In that epoch, climate fluctuations could have led to recurrent situations of population growth and then, with dwindling resources, crowding, and serious political competition. If small groups were annihilating each other at sufficiently high rates, a Darwinian group selection scenario that favored groups with greater numbers of morally upright, cooperative altruists, including men acting as warriors, could conceivably have been in play quite significantly.[16]

Here, however, we'll be concerned with what could have been a very powerful type of social selection, one that over evolutionary time could have been far more consistent in its effects than group selection if the latter were driven by warfare spurred by cyclical climatic downturns. This means that as far as moral evolution is concerned, we'll be much more interested in what happens *within* groups, than between groups. In that context, I'll be constructing a series of tentative and sometimes partly competing hypotheses about how human evolution, with major help from social selection, could have taken a turn in the direction of morality.

ANCESTRAL *PAN*'S HUNTING AND SHARING PATTERNS

As I've said, the CA cannot be designated a hunter because gorillas don't hunt. In contrast, Ancestral *Pan* was actively hunting—and sharing—some occasional small-game meat, and presently we'll be learning about some finer details from bonobos and chimpanzees that underlie this ancestral assumption. Since then, at some point weapons-bearing humans began to regularly and actively scavenge, hunt and share the carcasses of animals larger than themselves; we've been doing this for at least hundreds of thousands of years, and as recently as 15,000 years ago human foragers, certainly the great majority, lived in mobile bands whose males hunted large mammals enthusiastically and often.

It's possible that sometimes females were involved in this prehistoric hunting as well, for today there are a small handful of LPA bands that exhibit some female hunting,[17] and both chimpanzee females and in particular bonobo females are known to hunt actively. There surely were Pleistocene junctures when human populations were being badly decimated by cold or drought and consequently some bands were becoming too small to be efficient at hunting,[18] and at such times gaining extra hunters could have been useful as a way of increasing kill rates and reducing unwanted peaks and valleys in meat consumption. Thus, female hunters might have added to the general social flexibility that enabled Pleistocene humans to cope with a wide array of sometimes very sudden and often stressful environmental challenges.

LPA humans hunt large game as a dedicated activity that involves sophisticated systems of sharing. Chimpanzees and bonobos hunt casually and rather rarely, and in the absence of projectile weapons they go after smallish game.[19] Bonobos hunt even less than chimpanzees, and they tend to do so as individuals, whereas sometimes chimpanzee males seem to be quite collectively oriented in their pursuit of attractive fresh meat.[20] Both species appear to savor the fatty brains when it's time to eat and share, and at Gombe what I found fascinating was the limited way that up to a dozen or more excited apes shared these carcasses; chimpanzees and bonobos have similar modes of doing this. Normally, a higher-ranking individual who captured the carcass firmly possesses the meat afterward and takes a lion's portion, sharing with some of those who approach and beg for shares—but definitely not with others.

I'll never forget a chimpanzee hunt I witnessed my first year at Gombe, when a foraging party with numerous adult and adolescent males aggressively captured four colobus monkeys in less than half an hour and began eating them. As a novice fieldworker, I asked Jane Goodall afterward why the alpha male, Goblin, had stayed on the ground throughout the hunt instead of climbing into the trees and participating. With the benefit of over twenty years of field experience, Jane told me that Goblin was waiting for someone else to make a kill so that he could selfishly confiscate it, and her interpretation made a perfect fit with what I had described that afternoon in my field notes. With so many kills being made, the alpha male knew he had a sure feast in his near future, so this habitual bully conserved his energy and let an adolescent male catch his meat.

It's clear enough that meat is a prized food, for apes that are not given a share seldom leave the scene. They steadfastly bide their time hoping somehow to get some meat, and often enough these unsuccessful chimpanzee beggars become highly frustrated and prone to quarrel among themselves. In both bonobos and chimpanzees, the impression is one of gluttonously selfish possession, combined with rampant cronyism when it comes to limited sharing of this most precious of all foods. The overall sharing process seems to be shaped both by individual political power and by personal alliances.

TOLERATED THEFT OR SOCIALLY BONDED ALLIANCING?

Biologist Nicholas Blurton-Jones sees chimpanzees' meat-sharing as a kind of "tolerated theft."[21] This means that it's all about power, and the meat possessor isn't really being generous at all. Rather, he or she realizes that the other hungry apes could fight to take away the carcass, so it's best to share it with them—and thereby preempt their strike. However, in free-ranging chimpanzees I know of no record of a gang attack in which a stingy meat possessor was physically assaulted and dispossessed, even though often enough up to half or more of the apes present are being given no meat.

I have copublished a somewhat different theory, which comes from my having watched a fair number of hunts in the wild over a six-year research period, with eighteen months spent actively in the field.[22] Jessica Flack and I have suggested that the possessors usually share meat with just enough partners to gain the allies needed to firmly control the carcass, but not with many others who are anxiously signaling their desire for meat as they jockey for a better position in the begging line. Thus, rather than passively "tolerating theft," the meat-possessor, using his or her initial control of the carcass, is in fact quickly and actively buying a few allies—latent allies—who will help to balance power against the hungry nonrecipients. More generally, these allies are likely to serve as friendly political partners or as reciprocators in future hunts or sometimes as breeding partners.

Compared to tolerated theft, for chimpanzees in the wild the inferred political and emotional dynamics become rather different if we consider this balancing of power with socially bonded others.[23] I believe these dynamics may hold also for the less-well-studied bonobos, for *Pan paniscus* follows a similar pattern of sharing with a favored few and excluding others, again without fights breaking out between the possessor and the excluded. This "alliancing" theory is congruent with the fact that at other chimpanzee field sites there is further evidence of political alliances being actively involved in meat-sharing. For instance, at the Kibale field site in Uganda, certain pairs of males enter into productive meat partnerships; if one of them controls some meat, he will share with the other—as long as such reciprocation is being continued over time.[24] And at Tai Forest in West Africa, where cooperative hunting seems to be in play more than elsewhere, individuals who have allied to cooperate in killing a prey also cooperate in eating the meat, while nonparticipants are excluded.[25]

The distinction between tolerated theft and an alliancing approach to meat control is rather subtle, but it's important for this evolutionary analysis because we'll be focusing on sympathetic feelings that are involved in social bonding. Tolerated theft interpretations make possible the assumption that what appears to be "sharing" actually involves no element of perspective taking or generosity, but rather just a fearful concession to the potential power of others. In contrast, with an alliancing interpretation it makes sense that sharing with favored allies would involve some social bonding and hence, quite possibly, some ape version of sympathetic generosity that combines with political expediency.[26]

There's no reason, then, that the two theories couldn't be combined, for we might surmise that the possibility of a gang attack leads a meat possessor to share with just enough bonded allies to discourage active incursions by those who are excluded. But the bottom line is that the meat possessor is, in effect, using some of the meat to purchase allies, which implies positive social feelings as well as fear of an attack. This pattern of meat "bartering" sometimes can apply to securing sexual favors from females as well, as a special bonus,[27] and again an emotional kind of bonding seems likely.

When humans share large game, aside from certain tensions and sometimes some superficial squabbling, there's obvious community joy in participation—because meat is so deeply appreciated, because no one is left out, and because eating meat together is a splendid way to socialize commensally. In observing wild chimpanzees, I've always noticed that the sharing process itself appears to be extremely tense and hostile among the competing beggars—but at the same time both tense and *amicable* between the sharing partners and sometimes downright friendly. (This is only an impression.)

Field reports suggest that the same could hold true for bonobos, who also share very sizable fat-and-protein-rich fruit items as adults,[28] and it's worth noting that in both ape species, as with humans, mothers regularly share food with begging infants and—again—that such accommodation should involve substantial positive feelings. This adult meat-sharing likely involves some kind of a behavioral extension of strongly selected maternal generosity,[29] but ultimately, at the level of genes, the frequent sharing with nonkin must be explained through benefits of political alliancing or through some other compensatory mechanism that repays the loss of meat when it is less than abundant.

Of course, even when ape mothers share with their infants, it's difficult to *demonstrate* that sympathetic feelings are at work, and such interpretations are still more difficult with adults. But if we set aside the question of motives, it's clear that Ancestral *Pan*'s rather limited sharing patterns—based on what bonobos do at their rather rare meat feasts as a least common denominator—provided an important preadaptation in terms of behavioral potential. Thus, when archaic humans finally decided to turn from hunting small game (and probably aggressively scavenging a large carcass once in a while) to actively hunting sizable ungulates as a major and regular part of their subsistence, they already had something rather significant to work with in the meat-sharing department—even though the sharing pattern would have been involved with dominant possession and the favoring of cronies and therefore would have been quite lopsided.

HOW HUMANS FOUND A DIFFERENT WAY

In *The Hunting Apes,* primatologist Craig Stanford takes the position that, although cooperative hunting was an important development in human evolution, sharing the meat was even more important. In today's bands hunting and sharing are greatly elaborated by cultural practices and symbols,[30] which means that in maintaining customary systems of meat-sharing, the political power moves tend to be far more subtle than with Ancestral *Pan*. In this context I think that even though some foragers habitually argue about whether ongoing meat distributions are by the rules and fair, or grouse about their shares afterward,[31] underlying this cantankerousness are usually feelings of goodwill. These emotions help to enable the sociable process of equitably sharing a large carcass, for everyone will be sharing a food that is nutritionally useful to all—and is supremely delicious as well.

The existence of these positive feelings is attested to by dozens of rich ethnographic accounts of sharing and also by the fact that *serious* conflict, as opposed to squabbling, is all but absent from the scores of instances of meat-sharing I've covered so far in my extensive LPA hunter-gatherer survey. Some ambivalence is expectable, of course, because humans are so heavily given to egoism and nepotism. But even though people from different families may bicker a bit, and in some groups they may complain loudly to remind others of how the system's supposed to work,[32] I believe that even the habitual complainers appreciate the benefits of their sharing systems, are able to make them work quite efficiently in spite of a few rough edges, and, again, can take delight in eating their favorite food with others in the band.

That we became efficient, cooperative, equal-opportunity meat sharers was important to our overall evolutionary success, for this expanded our diet breadth and allowed us to exploit major new subsistence possibilities.[33] At some level I believe that archaic *Homo sapiens*, with its relatively large social brain, must have understood something about the importance of hunting cooperatively and about the advantages of sharing meat among the entire band. Today, egalitarian hunter-gatherers definitely seem to appreciate the advantages of having a band with more hunters, for obviously this means that the big carcasses to share will come in more often and hence there will be fewer hiatuses when people simply have no major meat to eat.

I think they might understand several fairly obvious long-term advantages of *equalized* meat-sharing, precisely because evening out consumption of this prized food provides nutritional benefits that lead to everyone's being more energetic and healthier, for people who live precariously close to nature are likely to have such insights. In any event, modern theories out of behavioral ecology reach exactly the same conclusions and show that these human patterns[34] are very much like those appearing on a purely instinctual basis among pure social carnivores, such as wolves or lions.

All social carnivores face the same logistical problems. Not only do they have to hunt as groups in order to keep the sizable and difficult kills coming in at least fairly regularly, but they also have to share these large kills *fairly* even-handedly if their diet is to sustain an energetically demanding occupation like team hunting, which works best if all the members of the hunting team are decently nourished on a continuous basis.[35] Dedicated social carnivores like wolves or lions are invariably hierarchical, and basically it is an evolved social structure that determines who gets how much meat. The simple fact that group members know when to dominate and when to submit prevents competitive conflict from getting out of hand, and with these pure meat eaters the challenge for natural selection has been to evolve mechanisms that ensure that some significant level of sharing takes place whenever meat is not very plentiful.

In the face of all this hierarchy, the sharing needs to be sufficient to adequately nourish the lower-ranking team members and also to ensure that the overall gains of group hunting are not lost in fighting over meat. The likely mechanism would be that selfish, aggressive, higher-ranking individuals are evolved at least to be *tolerant* when it comes to sharing a carcass with subordinates. Thus, even though the sharing may be far from equitable, the large carcasses they kill will be spread around better than with bonobos or chimpanzees with the much smaller animals they capture.

The technical name for this evening out of meat consumption is "variance reduction."[36] In humans alone, something really close to *equalized* meat intake is accomplished—but only with the assistance of symbols, cultural inventiveness, and the unusual capacity of our LPA groups to collectively control (or eliminate) powerful individuals who otherwise would dominate the meat scene. Because we deeply appreciate the benefits of sharing, and because we understand the politics involved, we can use various carrots and sticks to back up the relevant cultural institutions and customs once they've been invented. In this way hunter-gatherers see to it that their overall systems of sharing will generally work smoothly—that is, without frequent and costly serious conflict.[37]

Archaic types of *Homo sapiens* stuck around for almost half a million years, with a relatively static and, as far as we know, relatively unimaginative stone culture that nonetheless was up to the tough job of Pleistocene survival. These humans turned to the intensive hunting of large game only 250,000 years ago, and by 200,000 years ago they were becoming anatomically but not yet culturally modern, even though their technology was improving.

When these archaic humans took on large-game hunting, to reduce major fluctuations in their family-level meat consumption they were obliged to share carcasses as entire bands, rather than just as smaller social units—units that probably were based strongly on maternal, fraternal, and sororal kinship and possibly also on pair bonding by breeding partners and paternal kinship. The alternative, in the absence of long-term storage, was for the lucky hunter's social subunit or "family" to have a short-lived selfish feast when he brought in a big carcass a few times a year, probably sharing some with a few cronies, and otherwise to endure very lengthy meat famines as far as sizable ungulates like antelope or zebra were concerned.

The set of hypotheses I shall be proposing involve earlier humans living in bands, and they involve male competition over scarce commodities like large-game meat and breeding opportunities with available females and also competition for power[38] in its own right. But before I begin to put together a group of possible scenarios for moral origins, some additional background will be needed.

EVOLUTIONARY BACKSTORY IN THE HUMAN LINE

There are a big handful of early "hominid" species that may or may not have been directly in the human line—I say this even though when their discoverers publish in *Science* or *Nature,* they sometimes tend to imply, or even claim, that their particular species is definitely a human ancestor rather than merely some upright ape that went extinct as a side branch. All are bipedal in gait, their brains are about the size of present-day *Pan,* and generally for several million years they show no strong evidence of *regular* pursuit hunting. We may speculate that as descendants of Ancestral *Pan,* the well-known *Australopithecines* and their like at least were very likely to have hunted some small game, and with equal probability they were at least making some clever tools out of softer materials. However, if a given species was heading for extinction, it could have lost some of these ancestral traits.

It may have taken several million years for fashioning stone materials into tools to have been expressed robustly,[39] but just after 2 MYA we do see an archaeologically known upright ape species that did some scavenging and possibly some active hunting. One of these terrestrial apes may or may not have been our direct predecessor, but their skeletons continued to look quite apelike aside from being bipedal, and their brain size was only somewhat larger. If the later, somewhat larger-brained species that Louis Leakey optimistically designated as *Homo habilis* was our direct ancestor,[40] then it would seem that in our line we've had the use of manufactured stone tools for several million years now, and we've been using them for butchery for the same amount of time. Unfortunately, it seems possible that Leakey's "humans" belonged to a highly dimorphic lineage that went extinct or even that more than one species were involved. Thus, some experts[41] are reluctant to include *habilis* as a member of *Homo*.

The first fossil with a definite and undisputed claim to human ancestorhood is *Homo erectus*,[42] for some time a contemporary of these later upright apes. *Erectus* appeared before 1.8 MYA and is far more reminiscent of us than any living ape or any prior fossil. Tall and slender but very strong, its body was built for long-distance walking or running far more than for tree climbing, and its skull held a brain far larger than that of any ape—even though it was only about half the size of our own. Within a few hundred thousand years, the African version of *Homo erectus* was making beautifully fashioned Acheulian axes, a stone tool industry that lasted with few changes for over 1 million years. And these early African humans were also hunting more and more as, after spreading into Eurasia as an extremely successful species, they evolved in Africa into the still-larger-brained archaic *Homo sapiens* we've been discussing. That larger-brained species, after staying around successfully for several hundred thousand years, eventually turned to the active hunting of large ungulates and soon thereafter evolved into modern humans.

Archaic *Homo sapiens* had quite large brains, indeed, and their bodies were made on the same lanky plan as *Homo erectus* but were still more modern. Toward the end of their evolutionary career, these archaics finally gave up on the very static Acheulian stone tool tradition that *erectus* had invented and began to manufacture implements that were more imaginative. But culturally modern they were not. For instance, they were not yet creating distinctive local cultures in short order by quickly elaborating their stone tool and other technologies in ways that were highly inventive, as today's foragers do. Such cultural creativity, along with such things as self-adornment with seashells, the carving of phantasmagoric sculptures, cave painting, and the fashioning of musical instruments, was to come later with cultural modernity.[43]

As early as 400,000 BP, these archaic predecessors of ours were hunting with carefully fashioned wooden weapons.[44] Remarkably, several spears probably belonging to them were interred in anaerobic soil in Germany and preserved to the present. They are nicely made and well balanced for throwing, just like Olympic javelins, and they may well have been used for killing groups of wild horses—which means that even then the people involved sometimes had some sizable packages of meat to divide up. However, we don't know how often they were killing such meat, and the sparse archaeological evidence cannot support the idea that back then big game was a staple dietary item that they depended on regularly enough to require a new system of sharing.

By 250,000 BP, however, according to archaeologist Mary Stiner,[45] the evidence for large-game hunting as a serious and routine pursuit of archaic humans becomes overwhelming. Our African ancestors' subsistence continued to be heavily based on plant foods, but large animal carcasses were being relied upon as well, and animal flesh was no longer merely an occasional major treat. Actively pursued sizable ungulates like antelope were now a staple, and for the theoretical reasons given previously the acquisition and disposition of this important food had to be well integrated into a nomadic foraging lifestyle that previously, in all likelihood, had involved a primary dependency on staple plant foods, along with some small game, far more than on occasional large-game windfalls. That's our general evolutionary backstory as far as subsistence is concerned. But with respect to politics within groups, there's much more to say.

A COMPLETE NATURAL HISTORY OF GANG ATTACKS

We're particularly interested in subordinate coalitions and in how they became potent enough to all but neutralize the alpha role in a species that remained (and still remains) innately prone to form social hierarchies.[46] At 8 MYA behavioral phylogenetics has told us with a substantial degree of probability that the CA had a largely unexpressed *potential* for subordinate coalitions to attack dominators who rubbed the majority of their immediate social community the wrong way. With gorillas as the least common denominator, such ancestral rebellions are conservatively judged to have been either merely potential or quite rare. However, at 6 MYA the same conservative methodology tells us that in all likelihood Ancestral *Pan* not infrequently attacked disliked dominators to reduce their power, with the possible outcome of wounding or, more rarely, death. This punitive type of social selection had at least the occasional effect of significantly disadvantaging the genes of those bully types who were attacked—even though in general domination surely continued to pay off handsomely in feeding and mating contexts.[47]

In suggesting that such social selection could have been lethal, I'm thinking about what we know for certain about the two *Pan* species and also about what seems very likely. As we've seen, humans use capital punishment widely, but only once has a non-scientist actually *observed* a counterdominant, *group*-inflicted death *within* a social community in today's *Pan*. Primatologists in fact have described several cases of subordinate rebellions that certainly *seem* to have resulted in death—but technically a scientifically conservative ethologist could count them only as gang attacks leading to "disappearances."

At the Mahale field site in Tanzania, chimpanzee Ntologi, an aggressive former alpha male, was gang-attacked by members of his own community and was never seen again by the Japanese fieldworkers there.[48] As chimpanzee males can't safely immigrate into hostile neighboring groups, he likely died. Goblin, the alpha-male chimpanzee at Gombe when I was there, was gang-attacked so fiercely that he fled into exile, subsisting for months in peripheral areas where—if he'd been caught by an enemy patrol—he'd have been killed right on the spot.[49] Goblin in fact survived to live for many years as a socially accepted male in the group that had rejected him, but Ntologi was never seen again. It's possible that he died of his wounds or that he was caught by enemies; he might even have died in exile of natural causes, but in any event he was definitely put at lethal risk by being exiled, as was Goblin.

Similarly, a sizable group of wild bonobo females in Zaire attacked a male bully, biting severely at his digits, and after moving away with many serious wounds, he was never seen again.[50] He may have died of his wounds, but it's at least conceivable that he might have immigrated to another group, even though bonobo males very often show definite signs of hostility when two communities meet by chance.[51] Thus, he too can be counted as a "probable" with respect to dying of his wounds. In both species, then, it's likely that even in the absence of manufactured lethal weapons, serious gang attacks can lead to mortality and hence a major loss of fitness. And we already know about the use of capital punishment by human foragers.

Accordingly, we may assume that throughout human prehistory well-armed subordinate coalitions, if properly motivated, could gang-attack high-ranking group members and do them serious or lethal damage. We may also assume that when large-game hunting was added to the human subsistence pattern, this provided a new social stimulus that favored a definitive solution for the alpha-male problem—in case it hadn't been resolved previously.

A MAJOR WORKING HYPOTHESIS

In scientific research we sort out theories according to their "provability,"[52] their predictive power, their general plausibility, and, more generally, how satisfactory the explanations are that they generate. In attempting to build an explanation of moral origins, the best I will be able to do is to provide working hypotheses—some of which some scholars may see as being merely glorified hunches, while others may see them as highly worthwhile leads for future research.

The theory I'll be developing in this chapter is that the advent of our moral conscience came through the agency of a special type of natural selection, namely, social selection as this has been discussed in Chapter 3 and elsewhere. Basically, this involved the effects of human preferences, either in choosing others in useful partnerships or in coming down hard on disliked deviants.

Specifically, the first part of the moral origins argument is that in our human past when group punishment became severe and frequent, this significantly affected the human gene pool because punishment reduced the fitness of deviants. The second and less obvious part of the argument is that as the severity and cost of such punishment escalated, this created a selection pressure that favored individuals with better personal self-control. The instrument of this better social navigation and more effective self-restraint was the evolving conscience, and in trying to make this theory as specific as possible—and to try to place conscience origins in time—I will have to create some rather speculative arguments. These will concern how and why punishment in earlier human groups could have escalated to become a major force that shaped gene pools.

In the natural Garden of Eden scenario we are going to consider, I shall propose that when humans embarked on a new kind of subsistence pattern that was based heavily on hunting, this could have raised predictable social challenges, challenges that could have been met only by groups cracking down on individuals whose behavior threatened the efficient sharing of a very special type of food. The only way to test a specific theory like this one is in terms of its relative plausibility,[53] and fortunately there are some key facts that can be marshaled to give it support.

HUNTING AND THE ALPHA-MALE PROBLEM

The main obstacle to setting up an equitable sharing system for a band of five to six hunters and fifteen to thirty people in all would have been an ancestral type of alpha male, who predictably would act as a dominator prone to appropriate the meat of others and favor his kin and cronies; this prehistoric alpha problem was identified several decades ago by archaeologist Robert Whallon.[54] There's still a great deal of the alpha dominator in our genetic nature today,[55] and, if personally uninhibited and socially unrestrained, this would quickly translate into gross inequities in the meat consumption of today's hunter-gatherers. It also would translate into tendencies to turn meat into political power because the possessors would likely be using their extra meat to favor kin and political allies and mating partners, just as took place ancestrally. Thus, whenever earlier humans went up against whatever preexisting ancestral type of alpha-male system was in place, they would have been playing a zero-sum game as they ganged up to do battle with dominant individuals who loved meat and were used to controlling and hogging it.

In this competitive game there was a great deal at stake for the well-fed dominants, whose nutrition previously had been coming out so far ahead, just as there were high stakes for the previously undernourished subordinates who increasingly would have been motivated to rebel actively as large packages of meat became more important to their well-being. Conflicts most likely would have been inevitable, as a newly equalized, culturally based sharing system was being put in place. And ultimately it surely had to be kept in place by the threat of force, just as takes place in the egalitarian present.

TRYING A SPECIFIC MORAL ORIGINS HYPOTHESIS

Ancestral *Pan's* society was hierarchical, with alpha males. We may be quite sure about this. We may also be quite certain that by 45,000 BP humans had created decisively egalitarian orders—something that bonobos and chimpanzees obviously have not managed to carry nearly as far because, in spite of their subordinate rebellions, they still have alpha males, and in the case of bonobos, alpha females. Somehow, at some point, we humans *decisively* got rid of our alpha males and became egalitarian. It's logical that such a definitive step in the egalitarian direction was motivated by a rank-and-file envy over the perks of alpha bullies, which related to power, food, and sex. More basically, the issue was personal autonomy, and I'd suggest that Ancestral *Pan* had a strong distaste for being intimidated and bossed around.

For explaining how the human conscience originated, the initial scenario I'm developing provides what might be called a likely ecological key. The several hypotheses I'll be dealing with here are necessarily tentative, but of course new archaeological findings—as well as new competing hypotheses—may provide ways of testing them further. Furthermore, future developments in behavioral genetics could eventually enhance such investigations. However, for now we must view them as working hypotheses that are difficult to test aside from assessing their plausibility in contrast to other theories of moral origins, which we'll be meeting with in Chapter 12.

In its basics, the initial hypothesis isn't very complicated. It posits that a quarter of a million years ago, when relatively large-brained archaic humans took on large-game hunting as a major and regular occupation,[56] they would have needed to share large carcasses, and share them efficiently as described above, so that entire hunting teams could remain well nourished and vigorous. We've seen that if alpha-male behavior were flourishing, this would have been a serious obstacle to such sharing, and that the only possible solution to this problem—the only one that I can think of—was for subordinate coalitions to have taken care of this problem forcefully.

Ancestral *Pan*'s limited but significant subordinate rebellions provided the preadaptation, and this hypothesis assumes that archaic humans could have escalated such behavior to the point that, to definitively control their alpha problems, they would have developed some systematic type of collectivized and potentially lethal social control. The aim would have been to prevent high-ranking bullies from just naturally monopolizing large carcasses killed by band members, and thereby acting as free riders, when it was the undernourished others in the band who were doing much of the hard work in hunting.

As a result of this very likely conflict, those powerful individuals who were better able to restrain their potential aggressions would have had better reproductive success than those who didn't—and got themselves killed. Thus, the evolution of more effective personal self-control could have been selected strongly. This can be taken as the beginning phase of moral origins, because it would have led to the internalization of rules and the development of a self-judgmental sense of right and wrong.

We may at least consider some further details. As better self-inhibition became individually adaptive, it could have applied not only to bullies but also to others whose antisocial behavior obviously threatened efficient meat-sharing, those disposed to act as meat-cheaters who wanted to hide carcasses they'd killed, or thieves who wanted to stealthily take the shares of others. When these three types of "deviants" started being punished by their groups, the result would have been that those who better inhibited themselves from taking such dangerous free rides were gaining greater fitness. It's precisely because our consciences often can inhibit seriously deviant behavior, and save us from punishment, that this hypothesis could explain conscience origins.

That's the first hypothesis. It holds that in theory social selection in the form of concerted group punishment might have begun rather abruptly as active large-game hunting phased in, because the collective punishment of intimidators could have been intensified wholly or mainly as a *cultural* development. Such a development required no further biological evolution, because in the form of subordinate rebellions substantial preadaptations were already in place 6 million years previously. However, there are several potential wild cards that could modify this hypothesis.

First, it's possible that earlier on alpha-male systems had already been subject to some substantial attrition at the level of the genetic dispositions that support them, simply because we know that in all likelihood Ancestral *Pan*—and also its direct descendants in the human line—was so averse to being dominated. Long before the regular hunting of sizable ungulates began, these increasingly large-brained humans might have slowly become better able to use subordinate coalitions to reduce the power of resented alpha-male dominators, not only because ever since Ancestral *Pan* they had always strongly preferred personal autonomy over domination, but also, quite possibly, because subordinate males wanted a greater share of mating opportunities.

Thus, punishment sufficient to accomplish some evolution of better self-control could have begun much earlier. However, the basic moral origins hypothesis I'm developing here could stay the same, for when hunting arrived, more decisive group power moves still would have been needed to take out whichever would-be dominators were still impelled to act on their ability to intimidate—while favoring those with better self-control. Similar social selection would have applied to serious cheaters and thieves, who also threatened a reasonably equalized, nutritionally efficient apportionment of meat.

With this second scenario the evolution of egalitarianism could have begun quite gradually, through an interaction of genetic and cultural factors. We may speculate that the earliest likely major escalation of egalitarian behaviors would have been with *Homo erectus*, for this would have involved some complex political challenges, and this first really certain human had a significantly larger brain than its probable apelike predecessors. Would the sharing of sizable meat kills have been part of the picture? With this earlier human there was an archaeologically visible interest in very large game, but this may have been actively scavenged only once in a while, when opportunities arose.[57]

Otherwise, the exact role of meat in that early diet is more difficult to determine, aside from the fact that Ancestral *Pan*'s pattern of taking small game all but surely was continuing in the direct human line.[58] If in fact some medium-sized large game also was being hunted actively and regularly by *Homo erectus,* then a systematic and aggressive approach to better equalized meat-sharing might have been invented as early as around 1.8 MYA. I say this because unlike very large mammals, such as elephants, game like antelope-sized ungulates do not provide a surfeit of meat. As a result, efficient sharing becomes useful.

It's difficult to project such speculations further back in time, for it seems likely that the earlier terrestrial apes in the human line would have had more limited social brains, while overall the archaeological evidence is quite scarce. In this connection, the more ancient the field site, the more likely it becomes that fossilized bones of less enormous butchered prey might have decomposed to the point that cut marks made by human tools might not be discernible.

Such a very early meat-sharing hypothesis cannot be absolutely ruled out, but as we've seen the solid evidence we have today does point to 250,000 BP for the beginning of active and regular hunting of sizable but not enormous game. And even if for some reason antihierarchical subordinate coalitions had begun to significantly whittle away at alpha power as early as with *Homo erectus,* the later advent of intensive large-game hunting could have greatly accelerated this political process.

What we may say, with a high degree of certainty, is that Ancestral *Pan* was hierarchical—and that at some point along the way, and surely by 45,000 years ago when cultural modernity had phased in fully, humans had become decisively egalitarian. We can also say that this was because alpha types were being put down, or executed, if they failed to control themselves and restrain their own power moves. This takes place today, and it is difficult to imagine any other way that it could have been accomplished yesterday.

When large-game hunting did phase in, if it were to succeed—and if decisive egalitarianism were not already in place—some really severe sanctioning would have been necessary as the inadequately inhibited self-aggrandizing alphas—along with greedy thieves and cheaters—were attacked by forceful coalitions of group members who ganged up to control the meat as a highly valuable form of "community property."[59] It's logical that merely doing the job halfway would have resulted in very high levels of conflict over meat, and that therefore, after 250,000 BP, the only viable course for efficient meat distribution would have been to suppress alpha behavior definitively.

This evolutionary hypothesis attempts to base a theory of moral origins in human behavioral ecology and also in an assessment of earlier humans' social behavioral potential. It awaits the creation of plausible alternatives by other scholars—and further archaeological evidence—as the only likely means of scientific testing in the very near future. In a different sphere, it's difficult to predict how long it will be before behavioral genetics may provide further key information that could be useful in making behavioral and chronological assessments.

At this point, we're left with really three alternatives. One is that archaic humans had not progressed very far beyond ancestral behaviors in the matter of keeping down alphas and that large-game hunting led to radical political change and also to some severe initial conflict in putting down the poorly inhibited alphas. Another would be that before that, with earlier humans their coalitions would have partially reduced alpha power—in order to improve personal autonomy and probably also to increase the breeding opportunities of lower-ranking males—and that this would have made the transition to relying upon large game much easier. The third would be that decisive egalitarianism was already in place when such hunting began and that in fact this might actually have been a prerequisite for large-game hunting to succeed.

Further research findings will be needed if we are to choose scientifically among these three hypotheses, but all three of them deal with basic variables that are pertinent to the emergence of an evolutionary conscience. With respect to such conscience evolution, this could have been quite gradual or heavily punctuated, depending on the hypothesis and on how powerful social selection, in conjunction with group selection, would have been.

OFFING THE ALPHAS

When Whallon pointed out that alpha hegemony would have interfered with the egalitarian sharing process that goes so strongly with hunting and gathering today,[60] he did not present a hypothesis for how an equality-based political order could have supplanted the old hierarchical one. However, in Chapter 4 we looked rather carefully at contemporary egalitarian foragers and found one drastic answer—in the form of capital punishment. There we saw that quite often this serves as a means of eliminating free-riding problems connected with overly dominant behavior. Once serious hunting began, and large amounts of meat were arriving in camp on a sporadic basis, very likely meat thieves and meat-cheaters would have been lesser targets, while greedy alpha bullies would have received the brunt of serious social sanctioning, just as takes place today. I emphasize that more than half the executions visible in today's LPA ethnographic record for fifty societies were likely to substantially improve the chances that everyone could share equitably in the distribution and eating of large carcasses, and as we've seen in Table I, among LPA foragers it was definitely the bullies, far more than the two sneakier types of deviant, who were singled out as the main targets.

When archaic humans began to hunt, let's assume for a moment that they were not yet fully egalitarian. Because the necessarily forceful imposition of new, more effective meat-sharing customs on alphas who were used to getting their own way was likely to have been bloody, this would have driven up the rate of social selection in favor of better personal self-restraint. On the other hand, if subordinate rebellions had already become more frequent and more effective, and if the power positions of tyrannical alphas had already been partly undermined, an equitable and efficient sharing of meat would seem to have been easier to set in motion and, as I just suggested, the rate of social selection would have been more gradual.

To summarize here, we cannot rule out the possibility that earlier archaics might even have already become *fully* egalitarian before 250,000 BP, which could have paved the way for large-game hunting to flourish quickly and with far less conflict because efficiently equalized meat-sharing rules could have been so much easier to impose. In this case, conscience evolution and moral origins would have begun earlier, and social selection might have been motivated more in rank-and-file desires for more personal autonomy, or in rank-and-file males wanting more breeding opportunities, than in needs to widely share large-game carcasses. However, whenever it was that a serious reliance on large game began, this new development still could have raised the rates at which alphas were being gang-attacked, wounded, or killed by subordinates who wanted assured shares of a highly precious food that arrived so rarely and in such large packages. And this could have accelerated the rate of conscience evolution.

This theory is basically political in that I have tied this strong selection force closely to the advent of egalitarian social orders. These hypotheses provide a very large window during which punitive social selection could have operated to make us moral, and these social orders could have begun to develop at any time in the course of human evolution, really. However, for today's definitive kind of egalitarianism to have flourished, it would have been necessary for human social and political intelligence to become powerful enough for subordinates to decisively curb the alphas in their bands.

It was when such punishment really took off that I think the older, ancestral, fear-based mechanisms of self-control could have been supplemented by newly evolved traits that added up to moral origins. These included more sophisticated perspective taking, the internalization of rules that made for more efficient adjustments of individuals to the perils of living in a group that was prepared to kill serious deviants, a sense of shame, moralistic judgment of oneself and others, and a special type of symbolic communication in the form of gossip.

This may seem a rather banal explanation for something as exalted as moral origins, but I offer it until a better theory comes along.

MORE ON PREHISTORIC CAPITAL PUNISHMENT

Presently, I will offer some archaeological evidence about butchering methodologies, which at least suggests that at 400,000 BP human foragers were not yet fully egalitarian. First, however, let's look to the more recent past and return to the issue of capital punishment as an antihierarchical measure. In this connection, consider three examples of more recent Magdalenian cave art from Spain,[61] which probably date back to the time when these culturally modern foragers were adjusting to the arrival of the Holocene Epoch, with its more stable climates. What we see in one is a cluster of ten male archers who seem to be rejoicing in something they have just done as they expressively wave their bows in the air. Lying on the ground some yards away is an inert human male figure who looks almost like a porcupine,[62] with exactly ten arrows sticking in him. That's all we know for sure, but some speculation is possible.

First, ten archers suggests a band of perhaps forty, which would be a bit larger than average today, but well within the central tendencies already discussed. Elsewhere in Spain, two similar depictions show three and six archers, respectively, so the overall average would be about six, which seems to be right at the average for contemporary foragers—even though with such a small sample size, this is merely suggestive. Second, with the killings done unanimously and at short range, this would *appear* to be an instance of execution within the band, rather than a very lopsided act of killing between bands. We can't be sure, but the appearance of this event three times suggests that it could have been an execution scene similar to the "communal" one described by Richard Lee for the Bushmen, where a serial killer was "porcupined" by his group.[63] We'll meet with the vivid details in a later chapter.

This interpretation is bolstered if we reconsider the political dynamics of group executions and the two ways that present-day foragers see to it that the executioners, who are fulfilling an important public duty, are not themselves killed by the executed deviant's angry relatives. Briefly for now, the main way is to delegate the bad guy's close kinsman to kill him, which is clearly not the case here. More rarely, the entire group may act as one to take him out, which jibes perfectly with what's seen in these cave depictions. Some further speculation is possible if we consider statistically the causes for capital punishment presented in Chapter 4 in Table I.

These three cave depictions might have shown the executions of fear-inspiring sorcerers or some other kind of bully, but we can only speculate about this. They might also have been prisoners of war, although this seems less likely. But what these depictions do tell us is that culturally modern humans were capable of killing someone they did not like, by acting in groups.

Capital punishment episodes are relevant to the selection of genes precisely because the fitness of those being punished is so seriously disadvantaged. However, as we've seen ethnographic underreporting makes it difficult to come up with any precise annual rates for capital punishment among today's mobile foragers. What we do know, however, is that, whereas the *overall* homicide rates for today's LPA hunting bands actually may equal the rates in dangerous cities like Los Angeles,[64] only a small (but still significant) portion of these would be capital punishment because most killings simply grow out of one-shot conflicts over women. However, prehistorically when bands needed to hold their more determined alphas in check in order to share large game, capital punishment was likely a viable and sometimes necessary option in dealing with the harder cases, and this could have profoundly affected gene pools.

ARCHAEOLOGY AND THE EGALITARIAN TRANSITION

Here's a further and socially important piece of evidence on earlier meat-sharing that supports the idea that definitive, egalitarian control of large-game carcasses could have arrived about a quarter of a million years ago, rather than much earlier. Mary Stiner and two Israeli archaeological colleagues[65] carefully examined the cut marks on bones from a number of large carcasses butchered by earlier archaic humans in the Middle East, which at the time could be considered a geographic extension of Africa.[66] They discerned very different patterns at 400,000 BP, when acquiring large game still seems to have been a side occupation, as compared with 200,000 BP, when active pursuit of hooved prey had already been a mainstay for 50,000 years. Those 400,000-year-old cut marks were chaotic and varied, as might be expected if several people had been doing the butchering from a variety of angles with different tools and using a variety of personal cutting styles. This earlier human pattern seems fairly consistent with the chimpanzee-bonobo meat-eating scene, which is replete with competitive political dynamics and generally involves several individuals working a carcass at the same time—even though one individual is clearly in control. With these Middle Paleolithic archaic humans, the sharp stone flakes used for butchering could have made it important to share without much conflict, for if serious quarrels arose over meat, unlike today's two *Pan* species the butcherers already had quickly lethal weapons in their hands.

In contrast, at 200,000 BP the cut marks are those of a single individual assuming a single position to butcher the entire carcass. The potential implications are enormous, for this later pattern is quite reminiscent of what takes place with modern hunter-gatherers, where in effect the meat becomes a vigilant band's common property, to be widely shared in a systematic, culturally routinized fashion that averts serious conflict.[67] To avoid even tempting the more dominant individuals to make selfish inroads into this latter system, predictably the carcass is handed over to some "neutral" meat distributor who was uninvolved with the kill.[68] This practice sees to it, by custom, that a successful hunter does not egoistically control the meat.

The single-butcherer archaeological pattern at 200,000 BP certainly seems to be consistent with this modern practice, and obviously the kind of LPA sharing system we have been talking about would not work efficiently if selfish alpha individuals remained as little restrained as they were in the epoch of Ancestral *Pan*, or as they may have been when earlier archaic humans went at carcasses so individualistically 400,000 years ago. All of this at least fits with the idea that a decisive system of political egalitarianism needed to be imposed if earlier humans were to *regularly* eat large-game meat with good nutritional efficiency, and—again—do so without undue conflict when lethal weapons were available to all.

THE MAIN HYPOTHESIS

Here, then, is a thesis that fits any of the above scenarios. Whenever it was that human groups became militant about their egalitarianism, logically it became highly adaptive for a band's alpha types to very carefully hold their dominance tendencies in check—and I've suggested that this, along with similar effects on those prone to act as thieves or cheaters, could explain how humans acquired a conscience. We may assume that these selfish "deviants" were genetically variable in their capacity for self-control. We also may assume that when these would-be meat hogs began to be punished regularly and severely, for antisocially throwing their weight around, this would have brought on group conflict at much higher levels than are seen today when alpha suppression is quite well routinized in LPA bands. As a result, with groups basically winning out over individuals, strong social selection pressures would have been at work on the genotypes of those prone to lose.

Over time, the apelike, fear-based, ancestral version of personal self-control would have been augmented, as there appeared some kind of a protoconscience that no other animal was likely to evolve. That is my moral origins hypothesis. A question I shall not really attempt to answer is why bonobos and chimpanzees did not evolve in this same direction, given that they shared exactly the same major head starts. But the answer may lie largely in the fact that humans developed more complex social brains—or, possibly, that humans became dedicated large-game hunters.

HOW QUICKLY DID A CONSCIENCE EVOLVE?

Darwin believed all natural selection processes to be quite gradual and constant in doing their work, which was stimulated by gradual changes in the natural environment. However, a considered modern view is that sometimes such changes can be so radical, and sometimes so rapid, that what biologist Niles Eldredge called "punctuated equilibrium" takes place.[69] This means that rates of genetic change may accelerate markedly owing to swift and major changes in the physical environment.

With respect to conscience evolution, there's also a *social* environment to consider.[70] If after hunting began alpha power was still substantial and punitive sanctioning had to become intensive, then just by itself punitive social selection could have provided a new kind of punctuation. I say this not only because intensification of capital punishment would have affected fitness so profoundly, but also because preference-based social selection would have been so well "focused" by human intentions. By this I mean that this selection force would have been subject to purposeful adjustments as everyday social problems were being coped with and crises resolved. In effect, the large brains of humans could have led to persistent patterns of ecologically oriented political problem solving,[71] which over the long term could have had profound genetic consequences. In being heavily cultural, these problem-solving styles could have been adjusted quite quickly, as physical or social environments changed.

"Normal" natural selection processes require minimally about 1,000 generations (for humans, about 25,000 years would be the equivalent) to bring a new trait into existence. That's what Edward O. Wilson told the world in 1978 when he wrote *On Human Nature*.[72] Even if hunting large game was a relatively recent punctuation point, at which we may hypothesize that powerful social selection forces arose to purposefully favor individuals having superior self-control, there would have been plenty of time available for as profound a change as conscience origins to have taken place afterward, as humans were heading for cultural modernity. The time frame can be estimated rather precisely. Hunting became intensive a quarter of a million years ago, whereas cultural (and moral) modernity had arrived by 45,000 years ago or perhaps a bit earlier. Minimally, this would have allowed for 7–8,000 generations of gene selection to do their work in making us moral or in finishing that job, and with present knowledge I'm proposing that this was accomplished to a very significant degree by social selection, guided by highly consistent social preferences.

I must quickly say that with respect to our becoming altruistic, group selection, reciprocal altruism, and possibly all the other mechanisms discussed in Chapter 3 were likely to have been making contributions as well. But this is a question we will take up in Chapter 12.

What if definitive egalitarianism had *fully* arrived before large-game hunting phased in? Even though this would seem to be inconsistent with the mode of butchering that Stiner identified at 400,000 BP, with archaic *Homo sapiens* as the large-brained actors it is not difficult to imagine that egalitarianism might have arrived sometime between about 500,000 BP and 250,000 BP, while *Homo erectus* cannot be ruled out. However, whenever this did take place, we may assume that thieves, cheaters, and, especially, alphas were not going away quietly; that many were killed or otherwise disadvantaged along the way; and that the human capacity for self-control was advancing as a result of all of this drastic *social selection*. In fact, if we look at the not-infrequent social sanctioning of LPA foragers, with their rather frequent use of capital punishment, this process may still be in operation at the level of gene selection.

SEXUAL SELECTION AS BASIC THEORY

Why has social selection been given such priority here in comparison to the other models of selection I just mentioned? Let's consider Darwinian *sexual* selection as one very fundamental type of social selection.[73] Sexual selection operates strongly enough to support traits we must view as maladaptive and "exaggerated," such as peacock tails.

They can exist because evolved patterns of decisionmaking are channeling the selection process, giving it what might be called (at least metaphorically) a special "focus." In fact, sexual selection is very *well* focused by innately guided female choice. Think about all of those peahens preferring the more vigorous and fit peacocks, the ones who advertise their genetic superiority with the more magnificent multispotted tails. Consider, as well, the wide variety of energetically costly and often quite "exaggerated" male courtship displays in other species, described so well by Darwin, that cater to female choice,[74] and you'll realize that a preference-based type of selection can have such strong effects that otherwise patently quite maladaptive traits can be selected in spite of their heavy costs in fitness. Indeed, these traits are being kept in place solely because they provide such huge compensatory reproductive benefits to the chosen males during mating season. In this light, in trying to assess the obvious power of preference-based sexual selection keyed to mating displays, British geneticist Ronald Fisher long ago referred to interactive, "runaway effects."[75]

Punitive social selection is far from being identical. Common sense tells us right away, of course, that this can be a matter of being summarily executed—as opposed to receiving a satisfying sexual reward. Yet punishment, too, involves well-focused choices that also very directly affect reproductive success, for a dead would-be alpha male cannot breed. I hope that it will be possible some day to create mathematical models to take this special, group-actuated, punitive mode of selection into account more effectively. For now, however, I suggest that in helping a conscience to evolve, social selection could have acted quite strongly—strongly enough so it's likely that our brains could easily have been redesigned to accommodate conscience functions over a period of 7–8,000 generations, and possibly much more.

Keep in mind that punitive social selection continues to this day, both among the world's very few remaining viable foraging societies and in any other type of society that consistently punishes deviants who fail to control themselves. And keep in mind that among foragers there's far more to such negative social selection than capital punishment. I've emphasized this very dramatic type of group punishment because of the heavy costs imposed, but prehistorically lesser social sanctions in all likelihood also had ample generations to work with, and even though their immediate effects surely were weaker, they probably would have been employed much more frequently. For instance, the long-term effects of being ostracized or totally shunned by the band can mount up, and even just having a shamefully poor moral reputation can adversely affect an evolutionary actor's marital and other partnership prospects because of reputational liabilities.

UNIQUE PROPERTIES OF SOCIAL SELECTION

Donald T. Campbell told us that when biological evolution takes place, the basic mechanisms involve what he called blind variation-and-selective-retention;[76] he used all those hyphens to designate a single process that basically is devoid of purpose. This randomized process involves environmental pressures operating directly on the everyday phenotypes of variable individuals, with effects on gene pools and therefore on genotype. Here we've been concentrating on gang attacks and other group sanctions, and also as we're about to see in more detail, reputational consequences that act as agencies of selection. These originate in social groups rather than in natural environments, so they merit some special interest.

It's in this context that I've come up with a highly specific evolutionary hypothesis: *The killing, wounding, social exclusion, and social avoidance of aggressive (or cunning) deviants who do not rein in their predatory tendencies could have influenced earlier human gene pools, affected them so profoundly that a uniquely human conscience was able to evolve.*

This type of theory is not wholly novel, by any means. A small handful of other scholars have considered the possibility that punitive social selection can significantly affect gene pools, but they've done so in areas other than conscience. Not unsurprisingly, the original insight came from Darwin, even though the idea was not very well developed.

Then forty years ago, in 1971, biologist Robert Trivers clearly identified as a selection force the "moralistic aggression" that hunter-gatherers are capable of.[77] He suggested that when such aggression was turned against individuals who defaulted on reciprocating relationships, this would have reduced the frequencies of genes that made for cheating. I have included this basic idea in my model, even though the emphasis has been on bullies far more than on deceivers.

ALEXANDER'S CONTRIBUTION

In the late 1970s biologist Richard D. Alexander also began to discuss social selection in humans—at least partly in terms of group punishment of cheating, as Trivers did, but mainly positively, in the context of mating choice, with worthier males being favored by females.[78] Alexander's student Mary Jane West-Eberhard went on to suggest that Darwinian sexual selection was just one type of social selection,[79] and she exemplified other kinds of social selection using mainly insect examples. As a biologist she continues to work at the task of broadly defining social selection as a particular type of genetic selection that stems directly from the social situations or social preferences of individuals.

In 1987, in his important book *The Biology of Moral Systems,* Alexander elaborated his thoughts about how cooperative reciprocity in meat-sharing worked in hunter-gatherer bands and how the inherent altruistic traits might be selected. The main puzzle was the way hunter-gatherers shared meat without regard for who killed it, and the way their expectations were geared not to a "tit-for-tat" kind of payback, à la Trivers, but to what Alexander called a system of *indirect reciprocity*—an important concept that was introduced in Chapter 3. Briefly for now, this meant that others were being helped on the general assumption that "if I'm generous to someone today, *someone* will be generous to me in my time of need."

Alexander was trying to define the altruism paradox strictly in terms of the practices of hunter-gatherers, and in this indirect reciprocity he saw a major puzzle. Obviously, in such a generalized system of giving and taking, the opportunistic free riders who took much more often than they gave would come out ahead in their personal fitness, while, conversely, the altruists who did more giving would come out with a deficit. This applied not only to the sharing of meat, but also to other beneficial behaviors, such as helping nonkin who become injured or ill or otherwise incapacitated.

Explaining how such systems could evolve involved two types of social selection. In viewing indirect reciprocity as a kind of insurance system for all band members,[80] Alexander had this to say about the role of group punishment: "Obviously, various forms of punishment, including ostracism or social shunning, can . . . be applied to individuals repeatedly observed not to reciprocate adequately or follow whatever codes of conduct may exist."[81] Alexander went on to emphasize, even more, the importance of social status or "reputation" to reproductive success. Although his main emphasis was on good reputations and their importance to cooperation, at this stage of my own argument bad reputations—and their social and genetic consequences—are of special interest for understanding conscience origins. A substandard conscience can generate not only a substandard reputation, but active punishment as well.

PUNITIVE SOCIAL SELECTION

Bad reputations are confirmed as community members gossip privately about the behavior of others. When groups of archaic hunters were coming down hard on their alphas, along with the thieves and cheaters in their midst, they very likely had the language skills to keep track of the deviants' entire social histories, which meant they were able to make them pay not only for single transgressions but also for long-term patterns of malfeasance. Developing a superior conscience that could make similar cumulative calculations in adding up past social liabilities, and do so on an introspective basis that was private and accurate, helped the better equipped of these potential deviants to stay out of serious trouble. Thus, the genes involved in self-protective self-assessment and self-control could have been strongly supported by punitive social selection.

Genes for aggression would have been affected as well. In 1988, in analyzing capital punishment among nonliterate humans of various types, political anthropologist Keith Otterbein returned to Trivers's original insight about moralistic aggression against cheaters to suggest that, over time, capital punishment could also have modified human gene pools to make our species less aggressive because the more aggressive types would have had their reproductive success curtailed.[82] Since then, Richard Wrangham has discussed autodomestication in our species, partly in the context of group punishment's having made our overall genetic nature less violently aggressive, and partly in the context of skeletal changes that go with the reduction of aggressive traits.[83] Wrangham's major interest in capital punishment as a social agency of gene selection parallels my own, and our views have been mutually reinforcing.

In 1999, in *Hierarchy in the Forest,* I considered the social selection effects of antihierarchical humans ganging up against their more forceful and willful superiors not only in this "curtailment of aggressivity" context, but also in a context that brought me to the present hypothesis about conscience evolution and moral origins. Here's what I said, little imagining that I'd be writing a book on the subject: "With the reduced reproductive success of extreme upstart types, natural selection seems likely to have changed our political dispositions considerably. . . . This could have taken place through debilitation of aggressive responses, strengthening of inhibitory controls, or both."[84]

It's the second effect that we're focusing on in this chapter, for strengthening of inhibitory self-control amounts to conscience evolution, and it describes a key aspect of moral origins. We may never be able to guess about what form a protoconscience took at the point when more efficient types of self-control were beginning to develop. However, I have at least suggested that the social scene in which this evolutionary development took place can be plausibly reconstructed on the basis of what we can suggest about Ancestral *Pan's* behavior and what we know about today's foragers and the problems they continue to face in keeping their social predators under control.

If we look back to Ancestral *Pan,* the only very likely type of decisive "social control" we've been able to reconstruct is rebellious coalitionary attacks that could have, on rather rare occasions, resulted in wounding, exile, or death. If we look to today's foragers, it's usually when the entire band's social or economic welfare is threatened by an unrestrained deviant that moralistic social control by the entire band becomes decisive and coordinated, as well as severe and sometimes lethal. Today, this social selection continues to do two things as far as our gene pool is concerned. One is to reduce innate dispositions to bully or cheat. The other is to keep our conscience in place as a means of self-inhibiting antisocial deviance that can easily get us in trouble.

MORAL ORIGINS THEORIES, SCIENCE TO MYTH

Later archaic humans with their assumed protoconsciences might strike us as nonmoral chimpanzees or bonobos strike us today. On the other hand, if we could observe them intensively, or subject them to experiments, we might sense that they had at least acquired some rudimentary feelings of "right and wrong" concerning the rules their groups subscribed to. We may never know. However, once earlier humans began to strongly internalize group values so that they were beginning to be guided by an internalized sense of right and wrong, I think we might perceive them as moral beings—particularly if we could listen to them talking about one another and if a judgmental tone of voice was used. And when they began to experience what we know as shame feelings and began to blush with shame, there would have been no question about their moral status.

Having a conscience is obviously of enormous importance to human social life, yet scientists seldom try to explain its origins. A sophisticated but limited theory is that of a pair of German psychologists named Eckart and Renate Voland,[85] who make the interesting suggestion that a conscience evolved as a moralistic means for parents to get their innately self-interested children to pay back some of the investment made in parenting them. Obviously, this "parent-offspring conflict" thrust is far narrower than the social-selection-based hypothesis I've just proposed, which looks to how a conscience functions in human social life as a whole and how it might be "designed" to be adaptive. My hypothesis also is historical and considers our ecological past in connecting conscience evolution with changes in how humans exploited their natural environments. There are many other ahistorical approaches to explaining "moral origins" that likewise seem to stop short of seeing how moral behavior could fit into a more general evolutionary framework, and they will be considered in Chapter 12.

And then, of course, there are theology and myth. Here, some contradictions I sensed as a young boy going to Sunday school[86] will briefly color the discussion. Both the Old Testament and the Koran portray an idyllic Middle Eastern Garden of Eden with abundant food and with no environmental dangers—aside from a manipulative serpent that comes and goes as it pleases and that, though certainly depicted as a most evil entity, may partly endear itself to some as a believer in free will and the quest for knowledge. Perhaps we should also add a vindictive Jehovah to this environmental dangers list, for it appears that this Supreme Being believed in entrapment. I say this because Adam and Eve could have looked forward to a potential eternity bathed in shameless innocence, with the race they'd founded living free of both social competition and moral compunctions—had Jehovah not set his trap. He baited it with attractively knowledgeable fruit, and human curiosity, urged on by a fearlessly free-lancing serpent, made Eve and Adam walk right in and put an end to what might have been (a somewhat boring) Paradise.

The scientific Garden of Eden that I discovered much later in life was quite different. It was anthropologically situated in the African Pleistocene, which provided some great opportunities in life but also some potentially very dangerous climatic instabilities, frequent hunger and hardships, real poisonous serpents, including hyperaggressive black mambas, and hungry big cats prowling around at night. But the writers of the Old Testament certainly got it right when they implicitly likened their original pair of humans to morally innocent animals, emphasizing that their fall led to moralistic worries about sexual modesty and many other things. The biblical story may be pure allegory, but it's worth noting that the theme of human-animal differences, combined with moral origins, appears widely in the mythologies of nonliterate people. In such purely oral traditions, the question of how humans acquired a shameful sense of right and wrong is addressed so frequently that I would have to number the instances in the thousands, and these stories can be colorfully different—yet strikingly similar.

When I was a graduate student, I spent a summer conducting field investigations with a research team on the Navajo Reservation, and one of the things we investigated was *Ichaa*, or "Moth Sickness," a type of mental illness that in Navajo belief befell young people who committed incest. There was a myth, relevant to the work I was doing, that I still remember very well. It was collected originally in the 1930s from a traditional Navajo by an anthropologically gifted Franciscan missionary named Father Berard Haile,[87] and the informant had been raised by the hunting and gathering generation of Navajos who raided widely in the Southwest before their defeat and incarceration in the 1860s.

It seems that for Navajos the earlier forms of humanity were like insects, and therefore they were able to breed closely with one another in the absence of any incest rules. One of their clans even had an insect name—the Moth Clan—and very happily these earlier "humans" were able to marry within the family, where love is always the strongest. According to the Mothway myth, after actual humans arrived, they naïvely decided to abandon their custom of not allowing brothers and sisters to marry, planning instead to follow the example of their Moth Clan predecessors. These early people gathered at a high place on the top of a mesa and married all their children to each other, while others went down to the base of the cliff to prepare a huge bonfire around which all the happy brother-sister couples would dance the night away. But suddenly, like moths, the young people were inexorably attracted to the fire at the base of the cliff, and en masse they rushed over the edge, fell into the fire, and burned to death. The moral of the story: if you behave incestuously like a moth, you're breaking a rule of nature and immanent punishment will follow. It was thus that the earlier Navajo people first met with crime and punishment.

The beauty of a fable, be it a Moslem or Judeo-Christian Garden of Eden story or the Navajos' Mothway myth, lies in the storyteller's using just a few poignant events to explain something as profound and complicated as human moral origins. The basis for believing these stories is "faith," pure and simple. The beauty of a scientific story, such as the one I'm in the process of telling here, lies in the fact that the evidence is designed to be weighed and the theories are designed to be challenged and, where needed, modified. In making my case, I appeal not to faith but to anthropological data and insights, to the far-reaching logic of natural selection theory, to what we presently know about brain functions, to findings from primatology and archaeology, and to all the other modern knowledge we have at our disposal.

WHAT, EXACTLY, IS THIS CONSCIENCE WE'VE ACQUIRED?

Broadly, a conscience provides us with a social mirror. By continually glancing at it, we can keep track of shameful pitfalls that threaten our reputational status or proudly and virtuously chart our personal progress as group members in good standing. But more than intellectual self-knowledge is at stake, for as a practical matter we're continually trying to cope with our own powerful, well-evolved "appetites," which so often are likely to land us in trouble with our groups. These run all the way from dominance tendencies to material greed and sexuality, and expressing them antisocially can create serious practical problems in everyday life.

Minimally, both the prefrontal cortex and the paralimbic system are involved in the emotional reactions[88] that contribute to personal social strategizing and self-control. And when the effects of group punishment began to improve our capacities in these areas, it was individual differences in the relevant brain functions that punitive social selection was able to work on in terms of underlying genes. Of course, with this social type of selection, as with natural selection more generally, it was basically variation in the phenotype that selection processes acted upon directly.

Ultimately, the social preferences of groups were able to affect gene pools profoundly, and once we began to blush with shame, this surely meant that the evolution of conscientious self-control was well under way. The final result was a full-blown, sophisticated modern conscience, which helps us to make subtle decisions that involve balancing selfish interests in food, power, sex, or whatever against the need to maintain a decent personal moral reputation in society and to feel socially valuable as a person. The cognitive beauty of having such a conscience is that it directly facilitates making useful social decisions and avoiding negative social consequences. Its emotional beauty comes from the fact that we in effect bond with the values and rules of our groups, which means we can internalize our group's mores, judge ourselves as well as others, and, hopefully, end up with self-respect.

LIVING WITH OUR CONSCIENCES

In the 1800s, Darwin knew in general that moral capacities were the product of our brains, but today we're beginning to put some of the specific pieces together. Psychologist Jonathan Haidt has demonstrated that our initial moralistic reactions can be based heavily on emotions, perhaps even more so than upon intellectual understandings.[89] As I've defined the matter, our consciences afford us the capacity to look back into the social past or forward into the social future, weigh the consequences of our actions *with feeling,* and adjust our behavior accordingly.

This same modern conscience goes beyond automatic self-inhibition and social strategizing, for some of us use language to literally talk to ourselves as we try to define, and if we can to resolve, the thorny moral dilemmas we sometimes face. In a sense our inner life is oriented to choosing lessers of two evils, and this capacity has been tested academically using MRI scanning at the level of the hypotheticals mentioned earlier. Some rather nasty favorites of philosophers include a burning house from which you (an adult subject) may rescue, say, either your mother or your sister but not both, or a situation in which you can stop a runaway trolley car and save five lives but only if you push the fat guy off the bridge and arrest the trolley's progress by deliberately making him die.[90] As we'll see, Inuit-speakers in the Arctic do something quite similar with their *children* in real, everyday life.

Much more likely moral dilemmas, which many Americans actually face as children or as young adults, include whether to engage in the underaged or otherwise illegal use of a "substance," whether to shoplift, or when driving whether to "stretch" a red light when doing so will "safely" save some precious time. Later in life, with half of our marriages experiencing adultery, the moral dilemma can be whether to stray and hurt the one we love. And then there's the matter of fudging an income tax return, which for many isn't much of a dilemma at all, *morally* speaking, because "the government" is conveniently defined as an alien predator. Such dilemmas abound all around us, and it's our consciences that guide us in resolving them. But even if we do manage to resist a particular temptation, the dilemma may still be there in the form of a strong psychological ambivalence that is not fully resolved.

It's through the conscience that such dilemmas are identified as problematic, and it's the conscience that mediates such ambivalences. Individuals can be highly variable in how they cope with these and other, more serious moral dilemmas, for some of us tend to be impulsive initially, with our consciences becoming active afterward in a damage-control mode, whereas others remain perpetually tempted but holding back. And then at the extremes are those who have internalized their group's cultural prohibitions so well that they are barely tempted, and of course they contrast sharply with the unrestrained psychopaths at the other end of the spectrum, the weakly conscienced people who were discussed in Chapter 2.

There also are "immoralists" who seem to take delight in breaking rules just to be breaking them, and in some modern nations this last reaction seems to be incorporated into youth cultures. We also have full-blown, rule-based alternative criminal cultures, with gangs (and, unfortunately, prisons) as their special breeding grounds. On the other hand, we also have monastic cultures in which individuals take oaths to live in a morally superior way and perhaps, like Thomas Aquinas, find their consciences being overworked because their naturally based ambivalences are so strong and their moral standards are so high.

I shall not try to suggest how all of these complex reactions evolved, but it's apparent that we've moved well beyond the rather straightforward, mainly fear-based modes of self-restraint found in the African great apes, including even apes raised by highly moralistic humans. Morality involves this special kind of self-consciousness we call a conscience, which enables us to think about several important things at the same time. One is the rules we've internalized, and another is our immediate desires. And then there are our larger social objectives in life, such as gaining a good moral reputation and avoiding a bad one. A conscience is the mediator of all this, and it's well evolved to do so.

We may tend to think of our consciences mainly as psychological agencies of moralistic self-control, as Darwin did, but I've suggested that we may define the evolutionary conscience as having functions that are much broader. An opportunistic, "Alexandrian" conscience enables us to predict the social reactions of our peers and to calculate what we can get away with socially and still keep up a decent moral reputation.[91] It also allows us to decide that a particular transgression is just plain worth it—even assuming that public discovery is likely. And it lets saintly types like the Mother Teresas of this world strategize their behavior so as to maximize their reputations and perhaps gain thereby, even though often enough their main motives may be altruistic.

The optimal evolutionary conscience, then, is not a perfect and complete instrument of rule internalization that makes self-inhibition automatic and might work very nicely in an anthill. Among humans things are quite different because individuals who take society's rules too literally, and therefore are too inhibited to behave with some moral flexibility, will usually find themselves at a competitive disadvantage with respect to relative fitness. Rather, a fitness-optimizing conscience is one that permits some bending of lesser rules for personal advantage, even as it recognizes which rules should never be bent because doing so will bring dire personal results. This is an argument from adaptive design. But it also fits quite nicely with what can be learned ethnographically about LPA hunter-gatherers and with what we can learn in a Cartesian manner by peering into our own evolved psyches.

MORAL ORIGINS

We now have a hypothesis about how moral origins began, for in all probability gaining a self-regulating conscience was the first milepost in human moral evolution. Basically, we've moved from a wolflike or apelike "might is right," fear-based social order to one also based on internalizing rules and worrying about personal reputations. This was enough to make us unique in the animal kingdom, but the real clincher was blushing with shame—a mystery of natural selection that no scholar has begun to explain to date. In terms of evolutionary priority, it seems to me that the internalization of rules and values probably came first, as one basic function of the evolutionary conscience, and that blushing somehow became associated with these self-control functions afterward. But this is sheer speculation. If someday the genes involved with socially triggered facial flushing and other moral reactions can be identified, a more specific theory might be offered and a chronological hypothesis might even be possible.

Becoming moral did not mean that our typical temptations were about to go away as facts of human life. Rather, they became entwined with feelings of anticipatory shame that had the effect of automatically inhibiting antisocial behavior from the inside. Nor did becoming moral mean that ancestral-type fear motivations were going away. We still behave ourselves in part because we dread the moral outrage of our peers—or, today, police intervention as well.

What happened was that the nature of self-control was transformed in ways that no scientist could have predicted—even though in hindsight some useful preadaptations are fairly obvious. And once people were becoming moral, their consciences did more than guide and inhibit them. We moved from being a "dominance-obsessed" species that paid a lot of attention to the power of high-ranking others, to one that talked incessantly about the moral reputations of other group members, began to consciously define its more obvious social problems in terms of right and wrong, and as a routine matter began to deal *collectively* with the deviants in its bands. In important ways the dominance of the group was superseding the dominance of individuals; indeed, the well-known social tyranny of small groups has been well appreciated (and resented) by potential or actual deviants for at least tens of thousands of years.

As we began to create the kind of gossipy, socially conformist moral communities that Durkheim described so well, the sense of right and wrong afforded by a conscience was able to transform group social control. It was by achieving a moral consensus that people who were threatened by serious social predation could connect with others, and do this so strongly that they could reach a point of shared *moral outrage* that simply by its threat could deter many a potential deviant. In their actual expression, the resultant punitive actions led to effective and usually safe elimination of those whose conduct seriously threatened or injured the common good.

As our gene pools changed as a result, an increasingly moral social life offered new evolutionary possibilities, which have been experienced by no other species. And one of these possibilities was the rise of altruistic tendencies as hunter-gatherers began to deliberately harness the good they saw in human nature. The result was social selection of a very different type, that contributed significantly to our becoming a species noted for its altruism.

The Positive Side of Social Selection 7

ALEXANDER'S BRILLIANT IDEA

I've just hypothesized that moral origins began with the appearance of a self-regulating conscience, but this would only have started the process of moral evolution. Once we had something like a conscience, this opened the way for a very different type of social selection, one that affected our capacity as moral beings in a sphere very different from self-control. To understand its power, we must look, again, to Darwin.

When Charles Darwin came up with sexual selection theory, he considered this to be a special type of selection process, guided by the mating choices of females, that could support otherwise very costly maladaptive traits. Human altruism can also be viewed as a maladaptive trait; it is similarly kept in place because patterns of decisionmaking compensate the altruists and enable the trait to be selected.

We've already met with the idea of *indirect reciprocity,* which was first introduced by Richard D. Alexander in 1979.[1] Alexander had taken note of two ethnographic patterns. One was that human foragers' band-level cooperation was long term, not short term. The other was that these people weren't acting as careful bean counters when it came to being generous to others who were in need, which ruled out anything like reciprocal altruism. Indeed, even though the most intensive aid was reserved for close kin, just as Trivers said,[2] these foragers were also prone to help nonrelatives, and they did so substantially in the absence of any social contract that bound their unrelated beneficiaries to pay them back in kind. This fits our definition of altruism. According to Alexander, whoever had the means to help nonkin did so without very carefully adding up any past history of giving or taking, knowing that in the future whoever then happened to be in a position to give help would also do so.

This brings us again to the Golden Rule, which seems to be expounded in all human cultures be they recent and complex or ancient and "Paleolithic." Some form of this prosocial dictum is found in the ideology of every institutionalized religion,[3] and as a generalization it has found its way into certain formal philosophies of ethics, as with Kant.[4] The essence of this dictum seems to combine elements of altruism and personal self-interest because in part it's a way to convince *others* to behave more generously.[5] Practically speaking, this rule also might be seen as an implicit invitation for free riders to "take advantage," but it's a universal idea, nonetheless, and universal ideas are likely to be there because at some level natural selection, combined with persistent aspects of human cultural lifestyles, has favored them in shaping our species prehistorically.

Even though most people are quite unlikely to have read Kant, they intuitively appreciate the Golden Rule. Alexander recounts a conversation he overheard in which truckers were discussing stopping to give aid to colleagues who had trouble on the road.[6] The idea the truckers articulated was that you help me when I'm in need, and I'll help someone else when he's in need, and he'll help still another. Thus, you're contributing to an ongoing system that sees to it that people get help when they need it—but the payback is merely probabilistic. Essentially, you have to trust the overall system to work.

Alexander's overheard conversation jibes with the dinner-table and drinking practices of Montenegrin truck drivers in the former Yugoslavia when I was doing research there in the mid-1960s. At a local truck stop and inn, I used to watch over-the-road Montenegrin *šofirs* (chauffeurs) as they enjoyed hearty peasant meals supplemented with liberal shots of plum brandy. They'd be sitting at large tables of six to eight people, and when it was time to pay, every last Serb always got out his wallet with a small flourish—with the certain knowledge that one person would be paying for all. I'd be holding my breath, but each time, by some subtle dynamic, the ritualized decision was made. When I followed up with an ethnographic query in other quarters, I was told that this was an ingrained custom; indeed, these tribal Serbs considered northern European tourists, who they'd seen dividing up the tabs when they dined together, to be hideously "selfish." Their informal philosophy seemed to be "what goes around, comes around," a philosophy that Americans sometimes apply—and sometimes don't—when it's time to pay for dinner with friends.

The isolated mountain tribe where I was doing two years' field research at the time was almost five hours by foot or horseback from the main highway, with neither electricity nor running water, let alone a wheeled vehicle or a restaurant.[7] And there I discovered that this truck drivers' ritual had ancient roots. When people were gathering in social groups, it was always just one man who passed around his bag of contraband tobacco so that each person could empty out enough for one cigarette into his newsprint "rolling paper." As with the truckers, the donations were made in a spontaneous fashion with no hint of repayment, except for the assumption that next time *someone* in a position to do so would surely be coming forward because that was the custom.

To me as a foreign observer, this ritualistic act of tobacco-sharing always seemed to be deeply enjoyable to the parties involved, and in the tribe a similarly prosocial pattern of indirect reciprocity also was followed in several more practical and economically important areas. These mountain Serbs were pastoralists, and their modest flocks of sheep were subject to two types of catastrophe. One was from wolves, who for reasons known mainly to wolves sometimes enter a corral at night and then go on killing sprees that end with the annihilation of the entire herd. The other was lightning, which can easily take out a clustered herd up on the treeless mountain pastures used all summer. When either of these rare calamities took place, up to thirty or more households would each donate a sheep to the unfortunate, who was thereby made fortunate again because his herd was restored.

My entire tribe of just over 1,800 souls was made up of somewhere around 300 households and just over 50 different clans, so these donations were coming partly from close blood kin but mostly from people who were unrelated—neighbors or godfathers or in-laws from other settlements. This meant that much of the generosity had to be extrafamilial and therefore altruistic. These local tribal networks are sized similarly to small foraging bands, whose "insurance" systems follow similar principles of sharing, and cooperation with nonkin as well as kin.

In fact, these continuing practices of traditional Serbian pastoralists—and of modernizing Serbian truck drivers—may have evolved culturally from similar customs in much earlier times, when these people were nomadic foragers who shared their large game. However, it's equally likely that they were invented later, simply because humans are inclined to come up with systems based on indirect reciprocity whenever they need "insurance" against bad luck. Meat-sharing is just one in a long line of mutual-aid inventions of humans in groups, and one way to keep the system working is to remind others to do their part when it is time to reciprocate.

Another tribal instance of indirect reciprocity that involved nonkin arose in the form of a *moba,* which was called when a Montenegrin family was building a house and for some legitimate reason found itself short-handed, or when it had too much hay to safely harvest before rain came. The *moba* was a volunteer work group, and again a mix of relatives and nonrelatives did the giving. Again, the social network from which the volunteers were drawn would be about the same size as a twenty-to-thirty-person hunting band, and in the case I observed of house building, the *moba* spanned two days and involved almost two dozen helpers. The host family put on a big feed each day, but the donated labor amounted to far more than the value of the cheese, bread, meat, milk, tobacco, and plum brandy supplied by the hosts, and fewer than half of the helpers were blood kin. I emphasize that precise individual repayment in kind was not envisioned, even though in general future help was anticipated as a response to future special need. As with the response to herds getting wiped out, this fits perfectly with Alexander's description of indirect reciprocity in hunting bands.

I must add that there was a palpable air of good feeling during a *moba*. The helpers appeared to be happy in their generous role, and there was an atmosphere of jolly camaraderie as people working together joked, drank their host's plum brandy, and ate as one big group there to help. And as with hunter-gatherers' indirect reciprocity, these various services were provided without any thought of carefully counting the immediate beans: you helped out others in need because you could, and in general you expected them to help you in your hour of need. That ideology supported a system that served people's special shortfall problems, and they trusted the system to work.

The analogy to modern insurance systems may be striking, but with these indigenous systems of indirect reciprocity there were no fixed payments: the donations were basically voluntary even though social reputations were at stake. Furthermore, it was a face-to-face social community, rather than an impersonal insurance company acting on advice from actuaries, that decided who was eligible. Even though I participated in only two *moba*s, it was clear from the way people talked about this practice that the *selo* (the village, in Serbian) would know if, by custom, someone was genuinely qualified as a recipient.

In my particular Montenegrin tribe, the "village" was actually a widely scattered localized settlement, but it was clear what people meant. They were speaking about the gossiping kind of morally based "collective consciousness" that Durkheim characterized so aptly,[8] and I can guarantee that if someone tried to get help when this was inappropriate, tongues would wag, social reputations would suffer, and some or all of the people in the hoped-for network might well fail to participate.

The same sense of what is appropriate goes for hunter-gatherers; when a vigorous and dedicated hunter is injured, there's little doubt that the rest of the band will help him and his family because his need is all too apparent—and the help will be all the more generous because this productive citizen obviously isn't trying to take a free ride. This is well understood by everyone, as is the fact that helping him to recover will be useful because there'll be more meat for everyone. In the case of a conspicuously lazy man, he'll be seen as a freeloader and the band is likely to help much less. By the same token, a generous person will be helped more in an hour of need than one who is stingy.[9] But the basic system applies to anyone with a decent social standing.

Thus, because of what might be called "macro bean-counting," in which general past patterns are taken into account and factored into a system of indirect reciprocity, such systems, even though *somewhat* vulnerable to free riders who are moderately lazy, are resistant to really flagrant opportunism because people won't put up with it. However, there remains the ultimate evolutionary question of how such "unbalanced" systems can stay in existence. The models tell us that the altruists who are helping nonkin more than they are receiving help must be "compensated" in some way, or else they—meaning their genes—will go out of business. What we can be sure of is that somehow natural selection has managed to work its way around these problems, for surely humans have been sharing meat and otherwise helping others in an unbalanced fashion for at least 45,000 years.

EMPATHY AND ITS LIMITS

As Frans de Waal argues so eloquently, empathy is an important and too often underestimated element in the social mix that makes us human.[10] I've been using Darwin's term—sympathy—because it is less technical, but to me such feelings, which include an appreciation of how others are feeling and what their needs are, become apparent in descriptions of the pleasure hunter-gatherers take in sharing meat, even though sometimes there's some simultaneous grousing about the appropriateness of the shares. Similarly, the pleasure that Montenegrin Serbs took in helping out their neighbors was obvious enough, even though some of the *moba* participants surely were thinking some about their undone work at home. More generally, I believe that humans are innately prone to respond positively to engaging in helpful cooperation—as long as they feel a social bond with those they are helping, as long as the costs aren't too high, and as long as they feel that, long term, the system will be insuring them against really bad luck.

Altruism, sympathy, and empathy aside, I believe that nonliterate foragers in their bands also have good intuitive understandings of their systems of indirect reciprocity and how they work. My overall impression is that even though these systems seem to be free of any compulsive long-term bean-counting, in the distribution of large game some specific types of reckoning do occur, in several special contexts. For instance, larger families are routinely given larger shares because their needs are greater, and, as we've seen, when generously participating individuals fall on temporary hard times (be this through injuries such as broken bones or snakebite, or illness), reasonable adjustments will be made for them even though generally their close kin will be the primary source of aid. In addition, rather often the hunter who made the kill gets a somewhat larger share,[11] perhaps as an incentive to keep him at this arduous task.

Although hunter-gatherer generosity is sometimes depicted as being all but boundless, the generous feelings that help to motivate band-level systems of indirect reciprocity in response to individual needs are not without limits, and the same is true within families. For instance, considerable bean-counting takes place with respect to the tradeoffs that attend supporting the elderly. Old people may at times be quite useful, in providing wisdom or helping with childcare, but at other times they become a serious liability to the immediate family members who primarily support them. Hunter-gatherers set limits as to how much they'll invest in family members who are becoming so infirm they can no longer walk effectively, as this is a serious problem for nomads who must carry small children and needed paraphernalia with them when they frequently change campsites. Most readers will be familiar with Inuit family practices of putting old people out on the ice and letting them painlessly freeze to death,[12] and the logistics are obvious enough. An adult who cannot keep up will be a substantial or impossible burden trekking across snow and ice, and when it's time for triage, everyone in the band understands the situation, which is faced with sorrow.

I remember very well asking Kim Hill, an anthropologist who has spent years in the field working with South American nomadic foragers, about whether such practices also exist in tropical situations. He said they do, but in this case what happens is that someone will come quietly up behind them with a stone axe and all but painlessly "brain" them. My obvious shock faded when he added that if they were left alive, to "die peacefully" Eskimo-style, predators might eat them alive—while scavenging birds would first go for their eyes. They preferred a quick death and sometimes chose to suffocate by being buried alive.

Here's a final note on altruism and its limits that comes, again, from my own fieldwork with a modernizing "tribal" people who still basically had a subsistence economy. In my highland Serbian tribe one day an eighty-year-old woman I'd never seen before passed the front of my house walking quite briskly, complaining to herself out loud as she propelled herself along with two stout canes. I was told she'd been left with no kin at all (*nema nikoga,* "she has *no one*") and that she had to walk from one house to the next every single day because, although by (altruistic) custom people felt obliged to give her food and a night's lodging when she showed up, no one was willing to let her stay a second night for fear she'd become their dependent. Thus, under certain conditions the quality of mercy outside the family can be strained and strained severely—even as a system of indirect reciprocity in fact is working.

At the same time, however, people who are embedded in these systems believe in taking care of each other as long as this makes sense by the rules they've invented. And the care foragers take of others seems to be nicely geared to the mechanical power of the three basic selection forces we've been discussing, as these favor, respectively, egoism, nepotism, and, at the end of the line but still very important, altruism. Charity may begin at home, but in a weaker form it extends to others in the group. And sometimes, as with modern blood banks, it can even be extended to total strangers.

INTRODUCING SOCIAL SELECTION

In 1987 Alexander was considering two theories to explain the altruism inherent in human systems of indirect reciprocity. We've already considered his views on prehistoric group selection, and we've also touched upon what in his mind seemed to be the more useful theory—social selection of the type that has been called "selection by reputation."[13] Such selection includes mating advantages and much more, and this selection-by-reputation approach of Alexander's has stimulated some considerable further research, even though many biologists have remained enamored with reciprocal-altruism or, more recently, the one-shot, mutual-benefit models that were discussed in Chapter 3.

Alexander's original main premise was that a reputation for being altruistically generous might bring fitness benefits that could compensate the costs of such generosity if, when parents or individuals were making marriage choices, they were prone to favor "prospectives" who had a social track record of being unusually generous. For instance, a socially attractive, generous male may be more quickly accepted as a marriage partner, and then this prime early marriage gives him a fitness advantage compared to other males who have to marry later and not nearly as well. If this reputational benefit outweighs the costs of the generous behaviors that made for the superior reputation, such generosity can bring a net fitness benefit to the altruist and the altruism is nicely explained in terms of natural selection.

Alexander's insights about social selection in humans helped to stimulate "costly signaling" theory,[14] which can apply to any species that has bisexual reproduction leading to mate choice. Other animals obviously don't gossip about reputations, but sometimes in their mating patterns they may be genetically evolved to favor individuals who give off costly signals that correlate with being of high genetic quality as mates. If the male's reproductive gains (namely, being chosen) exceed the reproductive costs (paying for the otherwise maladaptive signal), such signaling can be a significant aid to fitness for both choosers and the chosen.[15]

The idea is that the cost of an unfakable signal of high quality, which again might involve unusually vigorous peacocks growing otherwise maladaptively outsized spotted tails, or a human hunter's putting more effort into hunting and providing more meat for others, will be compensated as such worthy individuals gain preference as mates—which of course strongly improves their reproductive success. And as we've seen, the females who are better evolved to choose them will also be coming out ahead because the attractively colorful (or productively generous) males they prefer will be unusually fit, which helps the females' fitness. Because there's an escalating interaction between the evolving, discernible signs of quality and the evolving preferences of females who do the choosing, this provides the possibility of the "runaway" selection process that was touched upon earlier.

Although runaway selection has been difficult to model mathematically, in theory the selection process involved could become quite powerful—particularly if natural selection is making other adjustments that are useful to fitness. For instance, a prime and unusually fit male peacock's huge tail is obviously a liability in terms of predator evasion, and in reducing such risks, these big tails have evolved to reach their zenith during mating season but shrink for the rest of the year.[16]

With such costly signaling, the potential problem of free riders remains. If signals indicating high quality as a breeding partner somehow can be faked by others who in fact lack the desirable but maladaptive traits, then the genuine costly traits will not be able to evolve because these free riders will be out in front. Thus, if a genetically inferior peacock could readily grow a tail as resplendent as a highly fit peacock, this sexual type of social selection wouldn't work, and the entire pattern of having camouflage-friendly drab peahens choosing the best decorated, colorful (and most fit) males wouldn't exist; both sexes would be drab, with coloration devoted entirely to camouflage.

The same is true of empathetic altruism in humans. If such generosity could be readily faked, then selection by altruistic reputation simply wouldn't work. However, in an intimate band of thirty that is constantly gossiping, it's difficult to fake *anything*. Some people may try, but few are likely to succeed.

THE SOCIAL BOOSTING OF ALTRUISM

Even though Alexander's assumptions about selection by reputation seem correct, for ethnographic examples in 1987 he was obliged to rely on just a few prominent studies of foragers like the Bushmen,[17] for broader surveys of the type that archaeologists Lawrence Keeley and Robert Kelly[18] began making in 1988 and 1995 were then a thing of the future. Fortunately, anthropologists like Frank Marlowe and Kim Hill are now working with really sizable general evolutionary databases for hunter-gatherers, and their analyses are helping scholars in other fields to consider the facts on a less "anecdotal" basis.[19]

The substantial, highly specialized hunter-gatherer database I am developing at the Goodall Research Center at the University of Southern California focuses just on social behavior of LPA foragers. When this systematic research on hunter-gatherer social behavior began over ten years ago, with substantial outside assistance from the John Templeton Foundation, the idea was to look for diversities and universals in moral behavior, conflict, and conflict resolution. The long-term objective was to categorize or "code" the social behaviors that were most relevant to evolutionary analyses of social ideology, social control, cooperation, and conflict so that interesting extant patterns might be discerned statistically and better prehistoric hypotheses developed.

Because we'll be examining these data in some detail, the coding system I developed needs some further explanation. For the past six years my research assistant has had memorized a five-page list of 232 varied and highly specific social coding categories, which range from "group member selected to assassinate culprit" to "sharing with kin" to "aid to nonrelatives favored." She patiently goes through thousands of pages of hunter-gatherer field reports to identify descriptive paragraphs that are relevant to each of these 232 coding categories, and then she summarizes each chunk of data individually.

Eventually, I hope that a searchable database can be made public to put all of this organized data at the electronic fingertips of scholars interested in social evolution, but this will require a large investment and for some time to come I will be obliged to analyze the data by hand, which is time consuming even though, for 50 out of a total of over 150 LPA societies, this vast information is now at least coded and summarized. If this coding weren't done, the time needed to ask and answer precise questions of such a huge amount of data would be quite prohibitive, and the treatment we're about to meet with would have been impracticable.

A few years ago, inspired by both the late Donald T. Campbell's psychological interest in altruism and prosocial preaching and Richard D. Alexander's biological thoughts on selection by reputation,[20] I decided it would be useful to investigate quantitatively the extent to which hunter-gatherers approve of generosity, especially extrafamilial generosity. The result has been that ten of the fifty LPA societies in this ever-growing coded sample have been used for intensive analysis here. These groups represent major world geographic regions and favor well-described hunter-gatherers who were relatively little affected by cultural contact before being studied. The idea was to see whether these "preaching" behaviors were widely mentioned for LPA foragers, and it turns out that both intrafamilial and extrafamilial generosity are unambiguously espoused by all ten groups.

This unanimity is of great interest. When Don Campbell and I taught a graduate seminar together at Northwestern in 1974, we discovered that among all six early civilizations, starting with Mesopotamia and Egypt, "official" preaching in favor of altruistic generosity was predictable and universal.[21] If hunter-gatherers did the same, and did so universally, then this would be a likely candidate for being a human universal, which would suggest, in turn, that such preaching was closely tied to human nature—and that it might be important to evolutionary analyses.

This recent analysis of ten foraging societies shows that generosity to nonkin is regularly mentioned indigenously as something that people in the band should practice, and there was no rocket science involved in assessing the coded materials. There are five coding categories that relate to generosity, and what was done, using whatever field reports existed for each society (there were between one and fourteen), was simply to count how many times people's being in favor of such generosity was mentioned by ethnographers who lived with and studied these bands. The very strong patterns seen in Table II (next page) are typical of today's LPA hunter-gatherers, so they can be projected backward in time for at least 45,000 years.

All ten societies had at least one such mention of extrafamilial generosity's being favored, but before we consider these numbers, we must keep in mind that many societies had only a few sources in the form of field reports; that for some societies, such as the Kalahari !Ko, multiple reports were published by the same ethnographer; and that some ethnographers are far more prone to focus on indigenous social attitudes than are others. Thus, the unanimity uncovered here is quite remarkable.

TABLE II ACTIVE FAVORING OF EXTRAFAMILIAL GENEROSITY*

SOCIAL IDEOLOGY	% of Societies	Total Cites	Andaman Islanders	W. Greenland Inuit	!Ko	Murngin	Netsilik Inuit	N.Alaska Inuit	Plateau Yumans	Polar Inuit	Tiwi	Yahgan
Geographical area			ASIA	ARC	AFR	AUS	ARC	ARC	NA	ARC	AUS	SA
# of sources			14	3	8	9	8	3	9	4	11	3
Aid to relatives favored	100%	61	2	4	1	5	15	6	4	21	1	1
Aid to nonrelatives favored	100%	65	1	5	11	3	26	1	1	14	1	1
Generosity or altruism favored	100%	232	13	13	24	21	59	12	8	45	4	23
Sharing favored	100%	312	16	30	23	19	92	32	8	57	3	23
Cooperation favored	100%	234	9	27	14	6	63	49	8	31	5	11

*This table was adapted from Boehm 2008b.

We see, for instance, that the Netsilik in central northern Canada had twenty-six mentions in favor of extrafamilial generosity in eight ethnographies, whereas the Yahgan at the tip of South America had one mention in only three ethnographies. Given that in spite of inconsistent ethnographic coverage these prosocial preachings are unanimous, even a sample of ten out of fifty coded societies provides a firm basis for suggesting that today and yesterday altruism has been, and was, actively promoted—and amplified—among mobile, egalitarian hunter-gatherers everywhere. It's also of interest that these "golden rule" preachings and statements are found not only in peaceful foraging societies, but also in highly warlike ones like the Andaman Islanders in Asia. Clearly, the roots of the preachings are ancient, and the central tendency is very strong, indeed.

It's worth emphasizing that giving nepotistic aid to kin also was favored unanimously as an ode to family values. But what's most important, here, is that aid to *nonkin* was explicitly advocated as a behavior that group members collectively favored and expected of individuals. Surely, such manipulative preaching was done for a practical purpose that we've met with already: it was to behaviorally amplify the sympathetically generous tendencies of group members.[22] The larger purpose was to improve the overall quality of a social and economic lifestyle that depended very heavily on cooperative indirect reciprocity.

The remaining items in this table, also unanimously subscribed to, demonstrate even more broadly that there's a predictable human concern with generosity because this is essential to cooperation and sharing. Thus, this social amplification of altruism would seem to be deliberate, well-focused, and probably universal.[23] Hunter-gatherers appreciate both cooperation and social harmony, and they understand that the general promotion of generosity will serve both of these causes.

HOW ALTRUISTS ARE COMPENSATED

We must look more deeply now into systems of indirect reciprocity to see exactly how altruists can be transformed genetically from losers into winners. In *The Biology of Moral Systems,* Alexander sees several ways that generous acts can be rewarded so that an altruist can be gaining more than is lost.[24] First, he mentions outstanding altruists being formally rewarded, as when a modern war hero receives a Congressional Medal of Honor. As long as the award is not posthumous, this individual is likely to reap direct social benefits that will provide fitness advantages that may offset his risks. In a foraging band there are no governments giving out awards, of course, but a parallel might involve a well-appreciated man or woman who at personal hazard has saved another band member from a predator or snake, or perhaps a well-respected chosen group leader where this responsible role exists. In the case of a group leader, there may be some modest costs of energy or time, or some risks in trying to manage conflict, but there also are likely to be gains from having an enhanced reputation.

Alexander proposes a second kind of popularity contest, in which individuals showing generosity in everyday life are favored as future associates in cooperation. This is the main basis for the selection by reputation we've been talking about, and it includes not only marrying to good advantage but also making other beneficial personal alliances, be they social, economic, or political. Thus, individuals with a track record of contributing generously in everyday life can be more attractive as future collaborators in a variety of partnership contexts, which means the resulting fitness advantages could be repaying what these altruists have lost by behaving so generously in the first place.

There's a third potential payoff for altruists. To the extent that these acts of generosity help one local group to flourish in competition with other local groups, Alexander seems to follow Darwin in thinking that altruistic genes might become better represented in future gene pools simply because groups with more altruists will prosper and grow in competition with other, less altruistic groups. The way he expresses this is that the altruists' coresident altruistic relatives and all their altruistic descendants will profit because they are part of a group that can grow faster than other groups,[25] and the wording suggests that he has both a kin selection and a group selection model in mind. But because the bands he is considering are now known to contain mostly unrelated families,[26] the group selection model may be more appropriate.

Alexander's first two paths to altruism are based on individual preferences that lead to selection by reputation, whereas ultimately the last is based on how groups compete as units. All are subject to the free-rider problem because a "hero" could lie about his exploits or a selfish person could try to fake being altruistic and might be chosen preferentially by a potential spouse, and in this way either of them might reap rewards that boost personal relative fitness and make it possible for these free riders to outcompete the altruists. Any substantial degree of cheating can undermine either social selection or group selection and thereby make altruistic traits individually maladaptive. But fortunately, there appears to have been an effective and rather distinctively human cure for this free-rider problem.

HOW SUPPRESSIBLE ARE FREE RIDERS?

In evolutionary science, working hypotheses lie somewhere in between "educated guesses" that are made on a relatively shoot-from-the-hip basis and well-controlled laboratory experiments, such as those (in physics) that demonstrated the speed of light. In Chapter 4 my strong working hypothesis was that as they became culturally modern, the members of LPA foraging bands were keeping careful watch on other members as they gossiped confidentially to keep intimates informed about who was behaving well or badly, and that they could eventually arrive at a group consensus on this basis—and act on it to punish an identified deviant.

As today, these band members had both the social insight to identify and even anticipate problems that threatened the welfare of everyone in the band, and the ability to deal with them directly and decisively before the good guys were individually taken advantage of—and before their cooperating bands were torn to shreds by conflicts started by the bad guys' behavior. For instance, if powerful deviants who tried to gain unfair shares of large game *seriously* threatened a customary system of meat acquisition and sharing, which was vitally useful to all group members, the band's response could be still more dire than the Mbuti Pygmies' obviously very angry reaction against Cephu, the arrogant meat-cheater who in effect stole from his group. Because bullying also is likely to stir stressful and disruptive conflict, group reactions to this forceful type of free riding would have been very well motivated, indeed.

Prehistorically, as culturally modern hunter-gatherers, people's goal would have been to protect themselves from the worst opportunists. As humans who like us generalized extremely well, these foragers surely understood two things. One was that over the long run predatory patterns of social deviance potentially threatened everybody, not just the current victim. The other was that there was security in numbers if they wished to use collective sanctions in coping with

TABLE III SOCIAL PREDATORS*

TYPE OF DEVIANCE	% of Societies	Total Cites	Andaman Islanders	W. Greenland Inuit	!Ko	Murngin	Netsilik Inuit	N.Alaska Inuit	Plateau Yumans	Polar Inuit	Tiwi	Yahgan
Geographical area			ASIA	ARC	AFR	AUS	ARC	ARC	NA	ARC	AUS	SA
# of sources			14	3	8	9	8	3	9	4	11	3
INTIMIDATORS												
Murder	100%	248	13	7	11	25	77	2	11	25	5	38
Sorcery or witchcraft	100%	122	4	10	2	25	36	1	7	10	2	4
Beating of someone	80%	79			11	16	8	2	10	15	3	12
Bullying	70%	12		1	4		1	3	1	1		1
DECEIVERS												
Stealing	100%	99	4	1	6	13	26	6	6	13	4	19
Failing to share	80%	34	1	2	4	3	9		1	7		4
Lying	60%	48	3		3		15		4	19		3
Cheating (general)	50%	24		1		3	7			12		1
Failing to cooperate	40%	10		1			4	3		2		
Cheating the group (as in meat-sharing)	30%	9					1			7		1
Cheating an individual	30%	9				2	3			4		

*The above figures are derived from the author's hunter-gather database.

the problem. What they didn't understand, and I emphasize this to make it clear that I am not reading enlightened and deliberate "purpose" into basic selection mechanisms, was that when such communal moralistic aggression was consistently and harshly targeted against the same types of deviants over thousands of generations, this would have a significant impact on gene pools.

Table III follows the (often overlapping) coding categories I use in my research to show, in descending order of frequency, some of the main types of punishable social predation mentioned for this same sample of ten LPA societies. Exploitation through dominant intimidation was statistically prominent, and it could be accomplished physically through murder or through administration of a beating, through the use of malicious sorcery, and through other forms of bullying. Exploitation through deception could be accomplished by failure to share and failure to cooperate, as well as by active thieving or lying or by deliberate cheating in several contexts. For a person disposed to social predation, a rich array of free-riding choices existed.

However, for humans the biological theorist's conception of intimidating or deceptive free riders automatically coming out way ahead of innately gullible and all but pathetically vulnerable altruists does not play out that way, for very often these opportunists can be readily identified by their altruistic peers and punished (with genetic consequences) in a truly wide variety of ways (see Table IV, next page). Thus, we must ask whether traits that make for seriously antisocial free riding—free riding that invites severe punishment—may often be far more costly to the would-be free riders than are the costs of being generous for the altruists they are genetically competing with. If so, for humans alone we have a possibly definitive solution for the genetic free-rider problem.

Table IV shows coded data for the same ten forager groups with respect to the sanctions employed, and again some of the coding categories overlap. Some types of sanctioning are underreported, especially because there is indigenous reticence about using capital punishment, but also because ethnographic reporting by its nature is spotty. Thus, with further information the apparent central tendencies represented by these findings would be still more robust; most likely, the majority of these social measures are either very widespread or universal among LPA foragers, so the central tendencies are pronounced.

All these sanctions contribute to punitive social selection, which takes place when entire groups develop strong negative preferences toward antisocial free riders—and act on these biases. The table shows that the system of punishment is quite flexible in its possibilities. Public opinion, facilitated by gossiping, always guides the band's decision process, and fear of gossip all by itself serves as a preemptive social deterrent because most people are so sensitive about their reputations.

Table IV Methods of Social Suppression*

MORAL SANCTIONING	% of Societies	Total Cites	Andaman Islanders	W. Greenland Inuit	!Ko	Murngin	Netsilik Inuit	N.Alaska Inuit	Plateau Yumans	Polar Inuit	Tiwi	Yahgan
Geographical area			ASIA	ARC	AFR	AUS	ARC	ARC	NA	ARC	AUS	SA
# of sources			14	3	8	9	8	3	9	4	11	3
ULTIMATE SANCTIONS												
Entire group kills culprit	70%	15	2	3	1	1	6			1		
Group member selected to assassinate culprit	60%	23	2	1		2	10		5	3		
Permanently expelled from group	40%	15			7		5	2				1
LESSER SANCTIONS												
GROUP OPINION												
Public opinion	100%	178	4	1	27	6	37	19	11	31	11	24
VERBAL												
Gossip (as private expression of public opinion)	90%	61		1	4	5	12	9	6	10	1	11
Ridicule	90%	72	2	4	4	1	28	4	4	21		3
Direct criticism by group or spokesman	80%	33		1	5	2	3	10		3	3	5
Group shaming	60%	40	2	3	2		15	5		12		
Other shaming	50%	9			1	1	3	1		3		
SOCIAL DISTANCING												
Spatial distancing (move or reorient domicile or camp)	100%	104	6	6	25	2	22	5	6	16	2	10
Group ostracism	70%	48		3	8		18	8	4	1		6
Social aloofness (reduced speaking)	70%	16			5		3	1	2	1	1	3
Tendency to avoid culprit	50%	11			3		2	1	2			3
Total shunning (total avoidance)	50%	6			1		1	1		1		2
Temporary expulsion from group	40%	7	1		1		3	2				
PHYSICAL NONLETHAL												
Nonlethal physical punishment	90%	93	7	6	11	2	21	3	13	21		7
Administration of blows	50%	22			4		7	1	1	9		

*The above figures are derived from the author's hunter-gather database.

Every one of the social control mechanisms seen in this table is mentioned in one superb ethnography by Asen Balikci,[27] who lived with and wrote about the Netsilik of central Canada, and also relied upon earlier data, collected right at the time of contact, when people were not yet reticent about their indigenous practice of capital punishment. The majority of these social control types were mentioned for over half of the LPA societies sampled, and the most prominent were social distancing, ridicule and shaming, expulsion from the group, physical punishment, and capital punishment. Thus, we may reconstruct yesterday's hunter-gatherers as being well equipped to identify free riders, suppress their behavior (as with Cephu), and, if these foragers couldn't intimidate the free riders enough to keep them under reasonably good control, get rid of them.

WHY DIDN'T GENETIC FREE RIDERS JUST GO AWAY?

Obviously, severe social punishment can heavily damage the genetic interests of deviants who would controversially put their own personal prerogatives ahead of group interests. However, even after thousands of generations of such punishment, there obviously are still some rather strong innate tendencies to take free rides, as indicated so eloquently in Table III and also by the hunter-gatherer capital punishment statistics we examined at the end of Chapter 4. If group punishment has been so damaging to free riders for thousands of generations, we certainly must ask how the genes they carry have managed to persist so strongly in our human gene pools.

The answer is unobvious but simple: to the extent that many potential free riders take note of such punishment, and use their consciences to restrain themselves and stay out of trouble, this keeps them alive and well because in effect they have been "defanged"—and therefore are not targets of social control even though by genetic metaphor their poison sacs remain intact.[28] The implications for the selection of altruistic traits are profound, for if free riders usually don't dare to *express* their predatory tendencies, their enormous competitive advantage over altruists largely goes away. And this means that as long as the altruists are being compensated by reputational benefits, the playing field can be close to level. Indeed, given that the more serious free riders may lose far more than they gain because of punishment, the field may be much better than leveled.

If harshly punitive social selection effectively intimidates most would-be free riders, conscience functions also slow down these deviants because values and rules they have internalized enable them to anticipate shame and loss of reputation. For mathematical modelers this means that behaviorally little-expressed free-rider genes can remain statistically salient in human gene pools—at the same time that the genes of altruists can also remain numerous as long as the altruists are somehow being compensated. Williams's well-respected models do not really account for this peculiarly human outcome.

In future research on humans, I believe this phenotypic suppression of free-riding behavior is something that any theorist trying to resolve the paradox of altruism must reckon with. Of course, I realize that Williams's elegant and aggressively promulgated models have captured many hearts and minds, and that there's a beautiful logic in straightforwardly equating free-riding tendencies (genotype) with actual free-riding behavior (phenotype). But people in small bands are so good at discouraging free riders at the level of phenotype that some major rethinking is in order about what is likely to be taking place at the level of genotype.

Social behaviors like capital punishment, banishment, ostracism, and avoidance as a partner in cooperation clearly have helped in significantly suppressing the frequencies of genes that favor predatory bullying or cheating, and over time this autodomestication surely has somewhat reduced and modified our innate potential for both types of predatory free riding.[29] However, in terms of explaining altruism, I believe that the still more important effect has been to frighten *potential* free riders so much that they'll desist from their depredations—even though their predatory inclinations are retained and passed on to offspring.

Add this all up, and what we have is a system of social control that can drastically reduce the genetic fitness of more driven free riders whose consciences can't keep these dangerous traits under control, but that allows the more "moderate," would-be free riders to control themselves in matters that would otherwise bring punishment and still express their competitive tendencies in ways that are socially acceptable. It's for this reason that free riders haven't just gone away, and this is reflected strongly in the tables. In fact, egalitarian human bands predictably have a few individuals who are unusually disposed to actively bully or cheat others in their community and pay the price. This is true after thousands of generations of social selection.

It was earlier types of social control that caused a conscience to evolve, and it's an evolved conscience that makes individuals so adept at this important type of self-inhibition. Yet today a fair number of risk-taking hunter-gatherers find themselves being executed, banished, ostracized, or shamed because the temptations to take free rides continue. They hope they can get away with it, but often they're punished. Much less conspicuous are the far greater number who hold back for fear of being sanctioned and therefore do no serious damage to the altruists in the group.

I cannot use my ethnographies to demonstrate all of this statistically, for it's very difficult to show why a behavior is absent or how prevalent it might be if it weren't for fear of being sanctioned or pangs of conscience. But after Cephu was so thoroughly humiliated, any other band members sharing his free rider's propensity to cheat in a net hunt surely would have been seriously deterred from actively expressing such predatory behavior, for fear of being discovered, confronted, shamed, and threatened with banishment by an extremely angry group. And the same obviously went for Cephu himself. Thus, even though his free-rider genes remained in the gene pool, at the level of phenotype Cephu's free-riding behavior very likely was curtailed. And the ill-gotten meat was confiscated, so he was in no way a winner, while his sanctioners were repaid for their modest policing efforts by eating that same ill-gotten meat.

OTHERWORLDLY SANCTIONING

There's one more means of free-rider suppression, which might even be seen as a special extension of the conscience. It comes in the form of supernatural sanctioning.[30] For instance, foragers often have food taboos, which I suspect began to evolve long ago because they provided a dramatic way of warning inexperienced or careless group members against eating poisonous edibles. This is just a guess.

Table V (next page) follows my coding categories in showing just those supernatural sanctions that pertain to *social* behavior, and they might be seen as an extension of the conscience because consciences provide people with a means of moralistic feedback that is totally private, rather than public. Likewise, imaginary supernatural entities can quietly track behavior and then privately judge and punish the offender, just as the conscience does.[31]

As a deterrent, these agencies obviously can make people who believe in them less prone to commit murder, incest, or various other antisocial acts, especially in situations where the social group is less likely to learn of the crime. And in an earlier study that focused just on supernatural sanctioning and was based on a more sizable sample of eighteen LPA foraging societies, I discovered that such sanctions were often directed against precisely the types of deviance that were conducive to free riding. Here, in Table V, I have used just the same ten societies as were used in Tables II, III, and IV, and nine of them report moralistic supernatural sanctioning.

In the previous, more detailed study, it was clear that supernatural sanctions often are involved with food taboos. In the area of morals, they help to suppress free riding. At the bullying end of the spectrum the suppressed behaviors include murder and sorcery, and at the devious end they include thieving, lying, cheating, and lazy shirking. Thus, imaginary "overseers," as well as real and vigilant social groups,

TABLE V MORALISTIC SUPERNATURAL SANCTIONING*

SUPERNATURAL SANCTIONING	% of Societies	Total Cites	Andaman Islanders	W. Greenland Inuit	!Ko	Murngin	Netsilik Inuit	N.Alaska Inuit	Plateau Yumans	Polar Inuit	Tiwi	Yahgan
Geographical area			ASIA	ARC	AFR	AUS	ARC	ARC	NA	ARC	AUS	SA
# of sources			14	3	8	9	8	3	9	4	11	3
Supernatural sanctions mentioned	90%	172	15	12	9	8	70	9		31	3	14
Belief that supernatural powers will punish transgressions	90%	167	12	12	9	10	67	9		31	2	14
Belief that individual deviance harms entire group	70%	41	2	3		2	24	3		6		1
Use of supernatural to manipulate deviants	50%	18		1	1	5	9			2		

This table was adapted from Boehm 2008b.

have been suppressing free-riding behavior in most (and possibly all) LPA-type groups ever since humans became culturally modern—and probably somewhat earlier, if we assume that these supernatural ways of thinking must have taken some time to evolve.

IMPORTANCE FOR HUMAN SOCIAL BIOLOGY

If we add up all the effects seen in Tables I through V, free riders are obviously a continuing problem. However, their suppression is multi-faceted and quite potent and begins initially with a conscience, which internalizes rules that favor cooperation and disfavor social predation. The conscience serves not only as an inhibitor, but also as an early warning system that helps to keep prudent individuals from being sanctioned. Such individuals gain a fearful anticipatory awareness that the band has a variety of effective and sometimes dangerous social tools with which to manipulate, punish, or kill those who transgress. Conscience also involves a recognition that people have moral reputations and that these reputations can have social consequences in everyday life. There's also a fear of supernatural retribution that can come even if others do not discover an individual's deviance. In addition, there are those constant positive ideological reminders that promote extrafamilial generosity. And finally it seems likely that many people behave themselves because they enjoy feeling positively about their own conduct.

As long as those gifted with free-rider genes hold themselves in check, they can contribute to a cooperative economic system, play the role of being a good citizen, and still carry—and transmit to their offspring—an above-average complement of free-riding genes. I hope that ultimately these findings, based on relevant statistical patterns for appropriate types of contemporary hunter-gatherers, will make others less prone to take mathematical modeling at its face value, and therefore less liable to jump to what may amount to hasty negative conclusions about the ultimate limits of human generosity.

I say this for the benefit of general readers who have encountered such doubts in various highly popular works, such as those of Richard Dawkins, Robert Wright, and Matt Ridley,[32] all of which to some degree take the standard negative stance that was so strongly promulgated by early sociobiologists like Michael Ghiselin.[33] But I also say this for the benefit of thousands of evolutionary scholars—be they biologists, psychologists, economists, or anthropologists—in hopes that increasingly they will be willing to look beyond the ever-popular kin selection, reciprocal-altruism, mutualism, and narrowly defined costly signaling paradigms in their consideration of human generosity.

Our distinctively human means of free-rider suppression are so effective that as LPA hunter-gatherers we have succeeded in transforming our living groups from ancestral hierarchical societies, in which the bullying type of free riding can be rampant, to egalitarian groups in which an individual actively expresses such tendencies only at high personal risk. Indeed, egalitarianism can stay in place only with the vigilant and active suppression of bullies, who as free riders could otherwise openly take what they wanted from others who were less selfish or less powerful.[34]

In this connection, I want to propose an evolutionary credo far more optimistic than sociobiologist Michael Ghiselin's skeptical "Scratch an altruist and watch a hypocrite bleed."[35] I do acknowledge that our human genetic nature is primarily egoistic, secondarily nepotistic, and only rather modestly likely to support acts of altruism, but the credo I favor would be "Scratch an altruist, and watch a vigilant and successful suppressor of free riders bleed. But watch out, for if you scratch him too hard, he and his group may retaliate and even kill you."

IS THERE A SECOND-ORDER FLY IN THE OINTMENT?

Group punishment is a crucial part of this suppression-of-free-riding scenario, and especially in evolutionary economics some scholars have raised the specter of "second-order free riders"[36] as a theoretical obstacle to groups being evolved to punish predatory deviants as I have shown human foragers in fact do today—and surely did yesterday.

The insights have come from experiments in which subjects who make unusually greedy offers to others may be punished even though the punisher, in refusing their low offer, comes out behind and therefore is paying a cost to punish. The second-order free-rider problem arises when one person abstains from punishing in order to let others pay the costs, a behavior that in real life would gain this free rider a genetic advantage. These insights are derived from formal game theory experiments that are explored mainly with college students,[37] but also at times out in the field with tribesmen and a few foragers,[38] all of whom have the opportunity to give up money in order to punish others who seem unduly selfish.

As these scholars define things, participation in group punishment itself is genetically altruistic because costs are paid to do so, which means that if free riders can hold back from punishment and thereby avoid investing the time and energy, and sometimes the risk, they can avoid paying costs that others are investing for the common good. This means that even as the group is punishing would-be aggressive free riders like malicious sorcerers or other bullies, or deceptive free riders like meat-cheaters, yet another type of free rider emerges: the one who stands aside to let others do the punishing—and thus cashes in on the rewards without paying any costs.[39] In theory, this should result in the advance of the genes of these free-riding nonpunishers and the decline of the genes of cost-paying punishers—to the point that genetic dispositions to join in group punishment would seriously decline and free-rider suppression as I have just described it might, in theory, just fade away.

However, if we move from mathematical models and experimental subjects to the kind of people who've evolved our genes for us, my database shows that everywhere hunter-gatherers do in fact readily punish their deviants—and that at any given time some individuals will be much more active than others in doing so and that some may refrain entirely. Furthermore, my colleague Polly Wiessner, who has been doing fieldwork with the LPA !Kung Bushmen for thirty years, has never seen or heard of punishment of those who fail to join with the group in sanctioning deviants. And so far in my survey of group punishment among fifty LPA hunter-gatherers (see Tables I and IV), the punishment of nonpunishers is never mentioned in the hundreds of ethnographies even though punishment does take place so regularly—and even though there are plenty of abstentions.

My own opinion is that these abstentions need have no relation to free-rider genes. For instance, in dealing with the Mbuti Pygmy Cephu's arrogant cheating, most members of the band actively shamed him, but members of the several households that were genealogically close to him were obviously staying to one side and appeared to be neutral.[40] I believe that this simply involves social factors that apply to everyone, not just to those disposed to be free riders. By this I mean that it is quite predictable—and understandable by the rest of the group—that the close relatives or associates of a deviant may choose to stand back and let others deal with him harshly.

Such abstention may give the appearance of classical free riding, then, but over the long run there's no such genetic effect because the structure of social role expectations explains these abstentions—without any need for free-rider modeling. One day I abstain because it's my brother who has been caught as a thief, but I don't actively support him either. Another time I actively participate in group sanctioning because it's not a close relative who is the deviant.

As another example of costs and benefits evening out over time, consider more specifically hunter-gatherers' use of capital punishment, as this was discussed in Chapter 4. We've seen that once in a long while the entire group may simultaneously mob a deviant,[41] usually a bully who goes around intimidating group members, and takes him out through collective action. This equalized sharing of effort and risk does happen to neatly preclude the second-order genetic free-rider problem, but the immediate motivation is something else entirely. What foragers are worried about is, in fact, revenge.[42] They know that if one of them were to kill a serious deviant, even if he was a truly bad guy his angry and grieving close relatives might well love him enough to engage in lethal retaliation. However, if most of the group participates simultaneously in his killing, there's no way for his relatives to target and revenge-slay the person who killed him.

More often, however, execution by forager groups involves "delegation" (see Table IV, "Group member selected to assassinate culprit"), and this second pattern seems to be very widespread and may well be universal. First, the group arrives at a consensus that the deviant must be eliminated, and then usually a close relative is asked to do him in. The cultural logic is impeccable. When a family has just lost one of its hunters, it would be unthinkable for them to double their loss by revenge-killing another family member—especially an upright citizen who was willing to kill his own kinsman for the good of everybody. Here, too, revenge is averted, but in this case one person is generously and responsibly assuming the risks of playing executioner. Again, it's a matter of structural position when the rest of the group delegates a kinsman to kill his own kin, so from the standpoint of gene selection, those who delegate him to do the dirty work are not acting as genetic free riders.

There remains the fact that often lesser degrees of social sanctioning are actively initiated by a few, with the general support of others, and that a cornered deviant might be dangerous. For instance, initially just one man took the lead in lashing out against the arrogant Cephu. However, this initial shamer had already learned that he had the rest of the band's strong backing, for Turnbull's psychologically rich description makes it clear that the others with him were equally incensed. He also knew that his guilty adversary, Cephu, understood that he was being criticized on behalf of the band, and that Cephu would be sufficiently intimidated that he would be unlikely to attack the lead criticizer. This group backing also was obvious when subsequently a mere boy insulted this arrogant cheater by not relinquishing his seat, which might have been quite risky under different circumstances.

Such group dynamics explain why the individual risks for those who lead a well-unified group majority in sanctioning are minimal—as long as the deviant's close allies are standing aside. In turn, these same political dynamics also help to explain why second-order free riding appears not to have been a serious obstacle to the earlier evolution of group punishment—and why, in real life, "defectors" from group punishment don't require punishment because things will even out in the evolutionary long run.

My conclusion is that no matter what takes place in experimental laboratories, evolutionary assumptions about the need to punish nonpunishers require further and critical consideration, for in everyday forager life the "group dynamics" are likely to be different from small-group game-playing contexts, as these have been scientifically contrived so far. Perhaps more of these group dynamics could be built into future experiments, taking into better account what hunter-gatherers actually do and were likely to have been doing in their culturally modern past. Meanwhile, we are left with the fact that classical LPA hunting-and-gathering communities do regularly punish deviance, that with them desisting from punishment is not viewed as a punishable deviant act, and that this version of social control has been in effect, and successful and powerful, for thousands of generations.

This conclusion brings in a further fact about these small foraging societies, namely, their political dynamics. These groups do not always stay united, for a morally ambiguous act of aggression can split a group, with the aggressor's kinsmen and allies siding strongly with him while the rest of the group proclaims his deviance. In such cases, the group is straightforwardly divided to a point that the two factions are likely simply to fission and go their separate ways. But this was not quite the case with Cephu's deviance, or with many other cases in which relatives merely stand aside and let the rest of the band do the active sanctioning. We must be careful to distinguish between somewhat-less-than-unanimous moral sanctioning—and outright conflict.

A PREHISTORIC DOUBLE WHAMMY

Building on the discussion in Chapter 3, I have radically broadened the scope of what is normally referred to as social selection as this is practiced by a variety of species.[43] For humans, I've now included Alexander's positive social selection through reputational payoffs, and I've added group punishment, which I began to think about as long ago as in 1982 as a selection mechanism.[44] Group punishment also leads to subsequent reputational disadvantages, so punishment and reputations are intertwined.

It was the combination of selection by reputation and active free-rider suppression that provided a "double whammy." By considering these two basic human types of social selection in combination, I'm proposing what amounts to a far more comprehensive version of social selection theory than is found in costly signaling explanations of mating advantages, which recently have been of great interest to evolutionary scholars who study other animals like birds and who study hunter-gatherers.[45] The comprehensive, "moralistic" social selection theory of altruism that I'm proposing here can now compete with theories based on group selection, reciprocal altruism, mutualism, and costly signaling, along with any new theories that others may come up with, as we continue to seek better ultimate explanations for the widespread human practice of extrafamilial generosity.

Obviously, this moral approach applies just to our own highly cultural species. Other animals can't build a consensus by gossiping about favorable reputations, even though costly signaling mechanisms may function analogically. And only a very few species gang up socially in coalitions to punish individuals in the same group who rub them the wrong way. If Ancestral *Pan* hadn't provided us with a modest but significant preadaptation in this direction, it's difficult to see how our species could have either developed a conscience, or substantially neutralized the bullying free riders in its midst to become as altruistic as we are today.

The solution for the altruism paradox I've offered here looks mainly to this comprehensive, human version of social selection for ultimate causation. This has involved ongoing suppression of free-rider behavior and also some significant past modification of the underlying genes—especially where the free-riding tendencies involved a bully's approach and could easily result in capital or other severe punishment. It also takes into account some hard to gauge but probably rather limited contributions from genetic group selection and also, phenotypically, some very potent contributions from a variety of cultural amplifiers that encourage extrafamilial generosity. I hope this combination of models will help to further explain the altruistic aspect of our common humanity, which contributes so profoundly to our quality of social life and its overall cooperative efficiency.

LATER MORAL EVOLUTION

Altruism is important to moral evolution for several reasons. One is that the sympathetic feelings that so often underlie altruistic acts are built into the conscience; this enables us to connect emotionally with the problems and needs of others as we act on the prosocial values we have automatically internalized at an early age in growing up.[46] Another is that so much of the content of our moral codes is oriented to amplifying the human potential for behaving prosocially, and it is sympathetic feelings that provide the potential in the first place.

I've argued that we do in fact possess innately altruistic traits ("scratch me—or scratch yourself—or scratch anyone but a serious psychopath—and see an altruist, not a hypocrite, bleed!"), and without such traits conscience functions would be quite different. Indeed, our moral life would be based mostly in shame feelings and fear of dire punishment, whereas the prosocial preaching that leads to the effective and often sympathetically based cooperation we've seen with hunter-gatherers would be absent because it couldn't work.[47]

A shameful conscience gives us a sense of right and wrong, but there's much more to moral life as we know it. The key ingredient of sympathetic feelings provides a motivational basis for much of our altruism, and this is an important element in systems of indirect reciprocity, for the participants are emotionally responsive to the needs of other individuals. Sensing the needs of others can lead us to spontaneously respond with generosity, and this, along with counting on future benefits from the generosity of others, makes the system work. Altruism, in short, is important.

Once people had become culturally (and morally) modern, the state of human affairs was basically that of the LPA hunter-gatherers I've described in their contemporary incarnations. To judge from today's behavior, our recent forbears were mainly egoists and secondarily nepotists, but as I've argued, they also were significantly altruistic in their genetic nature. The resulting extrafamilial generosity gave them something important to build upon culturally when they needed to cooperate with a specific vision of the common good in mind—as when a large carcass was there to be shared and serious conflict was to be avoided. These more recent hunter-gatherers had consciences just like ours, and in many ways their virtues, crimes, and punishments surely were very much like ours, as were their personal sense of what was shameful and their sense of good or bad reputations in others.

Large social brains enabled these people to see that the limited altruistic tendencies of group members could be socially reinforced for the common good. And basically it's because of these powerful brains that people invented and maintained the systems of indirect reciprocity that have served them so handsomely and so flexibly over the ages. This socially constructive aspect of human brainpower has helped to make our evolutionary career fully as distinctive as Darwin thought it to be, and we will be exploring this flexibility further in its social and ecological aspects. But without a significant degree of innate altruism and extrafamilially generous feelings to work with, I suspect that this remarkable evolutionary career would have gone in a very different direction, indeed.

8 Learning Morals Across the Generations

A CLOSER LOOK AT MORAL COMMUNITIES

Aside from the vivid but brief qualitative introduction to people from two African forager societies in Chapter 2, I've been trying to identify patterns of moral behavior through representative statistics wherever possible. To provide now a richer and more culturally distinct impression of what moral life is like in several of today's LPA moral communities, I'll be turning in the next several chapters to the words of some of the individuals concerned as I develop additional ideas about their moral communities and about how our capacities for social conformity and self-sacrificial generosity could have evolved in the Late Pleistocene.

We've already met with Colin Turnbull's vivid account of the meat-cheater Cephu's shaming by his group,[1] but the Mbuti Pygmies can't really qualify as an LPA foraging society because they have Bantu partners who are tribal farmers—obviously a Pleistocene impossibility—and they regularly trade wild meat for domesticated grain. I chose to quote from Turnbull because his description of a Pygmy moral community in action was truly exceptional.

As a major founder of sociology, Émile Durkheim himself[2] never did any fieldwork at all, and he made little reference to indigenous people as individuals. However, he certainly captured the collective side of social life and social control in these small moral communities when he described the tyranny of public opinion, which produces an often-fearful social conformity that goes with living in an intimate and potentially aggressive band of hunter-gatherers. Durkheim gained his insights vicariously from reading classical early ethnographies that described Australian Aborigines,[3] and he read them well. Most critiques of Durkheim's work have suggested that his "functionalist" image of small societies and their integration was seriously "beautified"—which surely it was in the absence of any serious emphasis on conflict[4]—but not that his reading of the ethnography and these social dynamics was incorrect.

Basically, my own hunter-gatherer insights, like Durkheim's, are vicarious. The only pure foragers I've studied personally are wild chimpanzees, which are of great help in doing evolutionary analysis but obviously lack a moral life. However, in the 1960s the isolated nonliterate Navajos I studied in the field had moved only a few generations from their foraging roots. They continued their egalitarian worldview very strongly,[5] with an emphasis on generosity and a condemnation of stinginess that were striking. The more traditional pastoralist Navajos I worked with were at least seminomadic and egalitarian, even if they were not "pure" foragers.

The quasi-tribal Serbian pastoralist-agriculturalists I lived with in an isolated mountain valley in Montenegro for several years were obviously of quite a different ethnographic type,[6] yet as we've seen, their systems of indirect reciprocity bore some signal similarities to those of hunter-gatherers, and even after a century and a half of living mostly under a despotic tribal "state," their egalitarian ethos was still quite evident. Spending two years in one small settlement of Upper Morača Tribe also allowed me the experience of living long term in a minuscule

"Durkheimian" moral community as a speaker of the native language, and this background, too, was invaluable.

TWO SUPERB EXEMPLARS

If I were obliged to single out just one LPA foraging society to exemplify the moral life of all such societies, I'd simply have to go with the one that was best described ethnographically. That said, I'd be hard pressed to choose between the Kalahari-dwelling Bushmen, including the !Kung, the !Ko, and the G/wi, and a pair of Inuit-speaking groups from central Canada called the Netsilik and the Utku.[7] These two sets of cultures have ethnographies that are exceptional in their portrayal of moral life, and fortunately I don't have to make a choice. We'll use them both in this chapter.

This rich ethnography needs some introduction. The loquacious !Kung have been vividly described by author Elizabeth Marshall Thomas and by professional anthropologists Richard Lee, Polly Wiessner, Pat Draper,[8] and by many others, one of whom, anthropologist Marjorie Shostak, recorded the life history of Nisa—a !Kung woman whose story is somewhat atypical because Nisa has an unusually non-monogamous love life.[9] She also seems quite concerned with issues of generosity and sharing for, as Nisa presents herself to Shostak, Nisa often appears to be not only quite stingy, but also equally demanding of generosity on the part of others.

As we'll see, these feelings of jealousy over food may have been exacerbated when Nisa was, with great resistance, being weaned. That possibility noted, the scrappy Nisa is far from being an immoralist in her own !Kung culture, and even though her autobiography may be somewhat atypical, it provides a special and even unique window into Bushman life that will augment our qualitative analysis well, here and in Chapter 10.

In central Canada, the Netsilik Eskimos had the major advantage of being studied not long after the time of contact by an anthropologically sophisticated Danish explorer named Knud Rasmussen, and of being restudied just as they were giving up their traditional nomadic way of life by Asen Balikci,[10] an ethnographer who also made a very fine and now classic series of films on these seal hunters. In addition, the Utku Inuit-speakers, who live in Back River and are regional neighbors of the Netsilik, were studied by my old friend Jean Briggs,[11] who left Massachusetts for the Arctic in the 1960s just as I was beginning my graduate studies. She was adopted into an Utku family, the family of the carefully aggressive Inuttiaq, and her intimate description of Utku emotions and morals is remarkable. So is the fact that eventually she found herself at the center of a moral crisis. In *Never in Anger* Briggs recounts how she was ostracized for months by members of her host culture, and later this will provide us with a unique glimpse into indigenous social control—from the perspective of a sensitive "deviant" from an alien culture who had seriously violated the Utku code with respect to emotional self-control.

Briggs's scientific interest was not in ostracism, and certainly not her own ostracism, but in how children were socialized. Her studies of the Utku and later another Inuit-speaking group,[12] far to the northeast in Cumberland Sound, provide an excellent idea of how Inuit societies lovingly nurture their young offspring while they are internalizing the group's rules and values, even as they also use hypothetical-situation moral dilemmas to pose cruel and stressful choices for these same small children.

Another reason I have chosen to treat the quite disparate Kalahari Bushmen and Arctic Inuit speakers in tandem is that these rich ethnographies usefully showcase commonalities in hunter-gatherer moral life that prevail in spite of obvious and enormous environmental differences. Specifically, the Bushmen live in a hot and seasonally quite arid environment, with the bulk of their calories coming from plant foods even though nutritionally (and culturally) hunting is very important to them, whereas the Inuit are obliged to eat mostly seal blubber and caribou meat (they prefer the blubber) along with seasonal fish, even though a few plant foods are available in the stomachs of herbivorous prey. These differences are profound. Yet, as we'll see, the moral feelings, styles of group sanctioning, and efforts to control conflicts are in many ways very similar. It all begins with having an evolved conscience, which facilitates the internalization of values and makes people think about both themselves and others in terms of virtuous right—and shameful wrong.

MAKING THE CASE FOR INTERNALIZATION

The process of absorbing the values and rules of a culture is subtle, and usually it's all but invisible to observers—but it is important nonetheless. In considering the individual internalization of societal rules, we need to reconsider here several names. One is sociological theorist Talcott Parsons,[13] who in portraying society functionally, as a system in equilibrium, followed Durkheim in seeing individuals as being deeply and all but automatically identified with their cultures and their group's rules. Parsons saw the individual internalization of values as an important element in providing cultural continuity for social groups.

Another name is that of the well-known economist Herbert Simon. Basically, in his terms being culturally "docile" means that an individual can and does readily take on *any* behavior offered by the culture. As we've seen in Chapter 3, Simon came up with the idea that being innately good at learning the culture is so individually adaptive that it could actually be "subsidizing" some altruism.[14] Economist Herb Gintis has advanced these concepts through a modern evolutionary economics approach that demonstrates the interaction of both genes and culture,[15] and he has tied Simon's piggybacking model directly to moral internalization.

In effect, both Parsons and Gintis are talking about one very basic conscience function, namely, the personal absorption of rules and values. And this leads to the conformist tendencies in humans that were discussed long ago by biologist Charles Waddington and more recently by psychologist Donald T. Campbell.[16] Nothing could be more important for perpetuating the moralized kind of social life that humans lead, and children begin to learn about these rules early in life.

Our qualitative portrait of LPA moral internalization will begin with some suggestive if indirect evidence for hunter-gatherers' rule-internalization—something that ethnographers in the field don't usually even think about because they already have their hands full describing just the main adult social and subsistence patterns they meet with. Child-rearing is where this all begins, and the good news is that at least a few hunter-gatherer ethnographers have focused on children and the process of moral socialization. Still more fortunately, two of the best happen to have studied children among Bushmen and Inuit speakers, respectively, with attention to the details of internalization.

With respect to moral socialization, LPA foragers have benefited mainly from Briggs's long-term, intensive studies of Eskimo children, and from Pat Draper's investigations with the Bushmen. In *laboratory* settings, however, the moral development of children in modern societies has in fact been studied intensively by various scholars over the past several decades,[17] and with great success, following the pioneering work of psychologist Jerome Kagan.[18] And because the responses studied in modern children are innate, they apply as well to hunter-gatherer children because we all share the same genes.

The further internalization that takes place in adults is still largely taken for granted by scholars, even though the universal moralistic preaching in favor of altruism that Campbell discussed for early civilizations suggests that this is a human universal.[19] I've already documented such adult-level preaching statistically for LPA foragers, but here some more direct and personal evidence of moral internalization will be useful.

For adults, probably the best one piece of evidence I'm aware of comes not from the usual ethnographic generalizations, which basically take internalization for granted, but from anthropologist Eleanor Leacock's active participation in an indigenous activity.[20] The North American Cree do not quite qualify as LPA foragers because they were so heavily involved in the fur trade before they were studied, but the story I'm about to quote strikes me as typical, and it provides a major hint about how deeply-internalized altruistic giving can be among hunters who are all alone, out in the field.

Leacock accompanies her informant Thomas on a hunting trip, and while they are far afield they encounter two men, known to them slightly, who are very hungry. Thomas gives these acquaintances all of his flour and lard, and Leacock's quoted description makes clear that Thomas spoke considerable English:

> This meant returning to the post sooner than he had planned, thereby reducing his possible catch of furs. I probed to see whether there was some slight annoyance or reluctance involved, or at least some expectation of a return at some later date. This was one of the very rare times Thomas lost patience with me, and he said with deep, if suppressed anger, "suppose now, not to give them flour, lard—just dead inside." More revealing than the incident itself were the finality of his tone and the inference of my utter inhumanity in raising questions about his action.[21]

Mention of a "dead" feeling suggests deep internalization, and because this was a matter of being generous to mere acquaintances, we may assume that the generosity was extrafamilial. In a similar Hadza case, James Stephenson, an adventurous New York landscape architect who travels to northern Tanzania to hunt with the Hadza for extended periods, reports that one time a hungry pair of strangers were met hunting far out in the bush and, again, the quite-costly sharing seemed to be automatic and deeply ingrained.[22] The Hadza are, in fact, LPA foragers, and I'm mentioning this anecdote, published by a nonethnographer, because it fits with what Leacock told us, and because normally such descriptions don't find their way into standard ethnographies. I wish they did.

In searching my ever-growing coded database for additional direct evidence of adult values internalization, I found nothing to closely parallel these two very revealing generosity anecdotes. However, just after the turn of the twentieth century Edward Westermarck, a Finnish sociologist who was a sensitive analyst of moral emotions and who made adept use of world ethnography,[23] spoke of internalization without using the word. He gave a number of examples from sedentary tribal societies of people identifying emotionally and deeply with the customs of their tribal groups, and he also gave one example from an LPA forager group, which I shall quote here: "Mr. Howitt once said to a young Australian native with whom he was speaking about the food prohibited during initiation, 'But if you were hungry and caught a female opossum, you might eat it if the old men were not here.' The youth replied, 'I could not do that; it would not be right'; and he could give no other reason other than that it would be wrong to disregard the customs of his people."[24]

These three anecdotes are at least suggestive. In fact, Leacock's and Stephenson's accounts have a ring of authenticity that speaks convincingly, to me, of extrafamilial generosity's being deeply ingrained in the Cree and the Hadza. My qualification to make such a judgment comes from the hundreds and surely thousands of hours I've spent reading other people's hunter-gatherer ethnographies and also from the time I've spent with Navajos and Serbs as nonforagers who still live in small, cooperative communities. I can only wish that such revealing case histories were available in all of the several hundred ethnographic reports I have covered.

With respect to Howitt's anecdote, which does not relate to generosity, my ethnographic intuitions tell me that he was not necessarily getting a full account of what was going on. With respect to the youth's breaking an important food taboo when cultural enforcers (the old men) were absent, there's another enforcement agency that might well have been operative. We've seen that supernatural sanctions are widely believed in by LPA foragers, and it's likely that clandestinely eating a proscribed "possum" would be noted by such forces. If so, in his mind he—or his entire group—might be visited with some kind of dire punishment. Belief in supernatural sanctioning can aid in the internalization of rules.

Of course, as a very different kind of evidence we have the universal group-level sharing patterns that are routinized with respect to large game. People engage in them by habit, and with little real conflict, precisely because they have internalized the values and rules involved. There's also the fact that even the most prolific hunters seem to *enjoy* participating in these systems—and basically seem to give up their carcasses without too much ambivalence even though once in a long while cheating obtains. Even more telling is that injured or disabled band members are assisted by unrelated band members through their meat contributions on the basis of a group ethos that calls for helping those in need. As we'll be seeing in Chapter 11, such contingent help is given in part on the basis of past generous behavior.

On the whole, internalization effects seem to be rather subtle, and they're quite difficult to separate from other agencies of motivation involved with meat-sharing, such as apprehension about seeming stingy, which is based on reputational concerns and in extreme cases on fear of active punishment. For instance, the Cree trapper Thomas surely knew that if he turned down the pair of hungry men, they might "bad-mouth" him to people he knew and thereby damage his reputation as a properly generous man. At the same time, his costly generosity might very well be mentioned when they arrived back in their camp, and through the exchange of favorable gossip he might gain in his public esteem in his own camp. But neither of these socially expedient personal considerations would account for the "dead" feeling he mentioned with such gravity. He obviously had absorbed his culture's values about sharing and in fact had internalized them so deeply that being selfish was unthinkable.

The depth of emotions that accompanies internalization has also been emphasized by Pat Draper, one of the earlier Bushman scholars. Draper's account of a young !Kung woman who felt herself to have been shamed helps us to understand how shame feelings are involved with the internalization of values and rules:

> When an individual runs afoul of some norm and the sentiment of the camp is against him, he reacts in a way that seems extreme to a Western observer. Further, the way the wrongdoer reacts to the frustration of criticism suggests that the social norms *are very well internalized* by the individual. For example, a young woman, N!uhka, about seventeen years old and unmarried, had insulted her father. Seventeen years of age is late to be still unmarried in this society, and her father often talked with her and with relatives about eligible men. She was rebellious and uninterested in the older men who were named. (She was also having a good time flirting with the youths in camp who were her age-mates but judged too young to make good husbands.) In a flippant way she cursed her father. He reprimanded her and immediately other tongues took up a shocked chorus. . . .
>
> N!uhka was furious but also shamed by the public outcry. Her reaction took this form: she grabbed her blanket, stomped out of the camp off to a lone tree about seventy yards from the circle of huts. There she sat all day, in the shade of the tree, with a blanket over her head and completely covering her body. This was full-scale Bushman sulk. She was angry but did not further release her anger apart from this gesture of withdrawal. She kept her anger inside, incidentally at some personal cost, for that day the temperature in the shade was 105 degrees Fahrenheit—without a blanket.[25]

REARING CHILDREN TO BE MORAL

My own interest in moral socialization goes back to the beginning of my life as an academic, and it involves what might be called an anthropological tragedy as far as my earlier professional career trajectory was concerned. For sedentary nonliterate people who are *tribal*, as opposed to foragers, the moral socialization of children has been a focus in a small group of earlier tribal or peasant ethnographies,[26] and I would have contributed to this body of work had it not been for a type of mishap that sometimes catches up with ethnographers as they try to do fieldwork under exotic and politically tricky circumstances.

In 1972, I completed a Ph.D. dissertation that was based on collecting 10,000-plus definitions of 256 morally based cue words, which I had elicited from forty Serbian friends and neighbors in Upper Morača Tribe.[27] In 1975, I returned to Montenegro to conduct similar interviews with children of different ages in order to study the stages at which these moral concepts became partially or fully articulated. Unfortunately, just as I was returning to the field, a colleague from a large American university was accused of seriously abusing his research privileges elsewhere in the former Yugoslavia, and I was forbidden to begin this fieldwork in Montenegro. This was a real shock, but the result was that I eventually turned my interests to the study of wild chimpanzees—a decision I have never regretted, even though I still wonder what I might have learned about the formative Serbian conscience and the internalization of values and rules.

LPA hunter-gatherers everywhere are deeply moralistic about their rules of conduct, which is a good *general* way of demonstrating that they have internalized the underlying values. But how, exactly, did N!uhka develop into a Kalahari person of such moral sensitivity? When laboratory scientists working with children in our own culture demonstrate experimentally when and how rule internalization takes place and a conscience begins to form,[28] I believe they are tapping into a universal aspect of moral life that is far more difficult to document ethnographically than experimentally.

At about two years of age our children in America not only begin to recognize themselves in mirrors, but they also start to blush with embarrassment and experience feelings of shame.[29] Couple these patterns with tendencies to help others in need, which appear in the same age range or earlier,[30] and we're extremely well evolved to become both extrafamilially generous and moral—as long as an appropriate cultural environment is present. These ingrained developmental windows are so predictable that surely LPA foragers are dealing with the same developmental potential.

We also may assume, however, that some cultural diversity exists in how their children are morally socialized. In this respect, in studying Inuit moral socialization, Briggs's role was that of careful describer, not experimenter, and a striking finding she made was that the Inuit go out of their way to force children to think very early about serious moral problems they'll be facing later in life. In fact, they do this very frequently, and in our eyes they do so rather cruelly, through stressful teasing.

What they do is to pose hypothetical moral dilemmas fully as nasty as the ones used by Harvard philosophers in their research on the responses of adults, whose MRIs they are monitoring. Two contrasting runaway-trolley dilemmas have been designed to produce different degrees of psychological stress that will light up different areas of the brain depending on whether the fat man has to be pushed off the bridge to his death actively or the subject merely is throwing a switch to stop the trolley and save five lives by sacrificing one.[31] The dilemmas studied by Briggs are totally natural, and in this context one of her research articles is titled "Why Don't You Kill Your Baby Brother? The Dynamics of Peace in Canadian Inuit Camps."[32] Confronting children shockingly with their own emotions, including antisocial emotions, takes place in various parts of the far-flung Arctic, so this is not just a local anomaly. Briggs says:

> These games are small exchanges, spontaneous in occurrence but highly stereotyped in form, between a child and one or more other people, who may be older children or adults of either sex. Sometimes the older person teases the child in some standardized way: "Where's your [absent] daddy?" "Whose child are you?" "Do you wrongly imagine you're lovable?" "Shall I adopt you?" "Shall I hit your nasty old mother?" At other times the game consists in tempting the child to engage in some disvalued behavior: "Don't

> tell your sister you have that candy; it's the last one; eat it all yourself." There are a great many such games; they are played all the time by everybody and a very high proportion of interactions with small children take this form. Most interesting is the fact that the games occur with only minor variations in Inuit groups that have no contact with one another, and their forms remain stable over generations. . . . By making conflicts salient to children, they help to create a sense of danger and, ultimately, commitment to the values.[33]

Commitment to values means internalization, and Briggs makes clear how this emotionally based internalization of values and rules of behavior prepares the Inuit child for a conformative social life as an adult:

> All these playful messages are delivered in a form that is extremely vivid, personal, and larger than life, and that makes it easy for children to see the problems posed for them by their contradictory feelings, to perceive emotionally the fatal consequences of a wrong choice of behavior, and to feel that since they have "bad" feelings, they are themselves vulnerable to the sanctions illustrated for them. To be sure, at one level the medium of the message is just a game, and therefore a cathartic way of coping with the contradictions that trouble us in real life. But at the same time, at another level, a doubt is engendered as to whether it really is a game, and the dangerous possibility that it might not be—the resulting fear—must make children try harder to conform than they might do if they did not feel themselves vulnerable to sanction.[34]

In Briggs's opinion internalization is about the interplay of conflicting emotions:

> Moreover, the values themselves become emotionally charged as a result of the threats that surround them. A child who is asked,

> "Why don't you die so I can have your shirt?" may start to value the shirt and keeping it, more; and on the other hand, s/he may come to place a high value on giving, because it is difficult to give away something one wants to keep. And when s/he gives—the shirt or something else—the recipient too will value the gift because s/he knows it was hard to make. Similarly, the child who is asked, "Why don't you kill your baby brother?" may love more strongly, and may value loving more, to compensate for the dislike s/he is made aware of feeling, along with his/her affection.[35]

This clinical analysis accords with something we know from common sense. Humans are set up to be ambivalent when it comes to many of the social choices we make.[36] On one side there will always be our usefully egoistic selfish tendencies, and on the other there will be our altruistic or generous impulses, which also can advance our fitness because altruism and sympathy are valued by our peers. The conscience helps us to resolve such dilemmas in ways that are socially acceptable, and these Inuit parents seem to be deliberately "exercising" the consciences of their children to make morally socialized adults out of them. This was particularly the case with the Inuit Briggs studied in Cumberland Sound.

Deliberately and stressfully subjecting children to nasty hypothetical dilemmas is not universal among foraging nomads, but as we'll see with Nisa, everyday life also creates real moral dilemmas that can involve Kalahari children similarly. There's also Briggs's later, highly detailed case study of Chubby Maata,[37] a three-year-old Inuit child in the eastern Arctic, which involved six months of intensive observation. In this distant settlement morally instructive "teasing" was a routine and frequent aspect of socialization, more so than with the Utku.

Such well-focused studies are lacking for other LPA hunter-gatherers, but this is not because hunter-gatherer childhood has been totally neglected. Recently, a large edited academic volume, *Hunter-Gatherer Childhoods,*[38] came out showcasing a wide variety of studies. However, after a century of serious neglect of moral socialization, these more recent interests have been in areas such as the age at which children begin to forage, what patterns of breastfeeding and weaning prevail, who act as substitute parents, maternal availability, children's play, or how children conceive of death. These are all useful and fascinating topics, but stages of moral development or, more specifically, the internalization of rules and values remains to be further investigated in its own right.

Melvin Konner, who is both an anthropologist and a medical doctor, has written a still heftier recent volume on *The Evolution of Childhood*,[39] and because he is a member of the sizable cadre of scholars who early on studied the !Kung Bushmen, he uses hunter-gatherer information where it is available. He discusses Jean Piaget's pioneering work on how Swiss children go through stages in understanding rules, and he also considers physical punishment. His conclusion is that, in general, people who are egalitarian use punishment far less, whereas people who are hierarchical, or who practice warfare, use it more. The first definitely applies to LPA foragers, for they're always egalitarian; with them a warfare level of intergroup conflict seems to have been highly optional, at least under recent Holocene conditions. Thus, when children of most egalitarian foragers internalize rules, this is likely to take place mostly through gentle, firm guidance, rather than by anything like regular or severe use of the rod.

We must assume, in the absence of plentiful data, that foragers in different culture areas are quite diverse in some aspects of their child-rearing strategies, and later in this chapter we'll see how !Kung parents handle a real-life weaning problem. But at the same time, I emphasize that adults' socialization of their children will always be geared to mesh with the aforementioned inborn stages of moral development and the rule internalization they make possible.

Here are some educated guesses, then, about how rules become internalized among all LPA foragers. First, as children respond to their caretakers, they spontaneously come to understand that certain behaviors are approved and others disapproved. Physical punishment may be unnecessary, but sometimes, as we'll see with the !Kung, it's used when reasoning, scaring, and shaming won't do the job. Second, caretakers may deliberately inculcate rules and values, keying their manipulations to the child's readiness to learn. And because infants are so early in developing spontaneous tendencies to help others in need, in theory teaching them to share generously could show major results from quite an early age. However, in our own culture—in spite of this early innate readiness, which in terms of altruistic perspective taking leads to simple acts of spontaneous helpfulness—Konner tells us that socially at age three, children tend to hoard their possessions; only at age five do they become much more prone to share.[40] It would be interesting to see if the timing is different among foragers, whose special needs to share widely in everyday life may well make their socialization practices differ from ours.

NISA'S CRUEL SITUATION

Comparable in important ways to the Inuit-wide, teasing style of moral confrontation is a real-life problem of triage that involved Nisa, as a Kalahari child who was well past being an infant. Nisa's active struggle to avoid being weaned as the birth of a younger sibling approaches is all but epic, and it provides, I think, a major clue as to how the !Kung Bushmen introduce their children to social control—and to how rules of conduct are internalized in that culture. And eventually this involved her in a much larger dilemma that was faced by her mother.

As a still-nursing child Nisa was very persistent as weaning began, and as a result she seemed to be highly ambivalent about her future sibling:[41]

> I remember when my mother was pregnant with Kumsa. I was still small and I asked, "Mommy, that baby inside you . . . when that baby is born, will it come out from your belly button? Will the baby grow and grow until Daddy breaks open your stomach with a knife and takes my little sibling out?" She said, "No, it

won't come out that way. When you give birth, a baby comes from here," and she pointed to her genitals. Then she said, "And after he is born, you can carry your little sibling around." I said, "Yes, I'll carry him!"

Later, I asked, "Won't you help me and let me nurse?" She said, "You can't nurse any longer. If you do, you'll die." I left her and went and played by myself for a while. When I came back, I asked to nurse again but she still wouldn't let me. She took some paste made from the dch'a root and rubbed it on her nipple. When I tasted it, I told her it was bitter.

It was at this point that physical punishment entered the picture:

When mother was pregnant with Kumsa, I was always crying. I *wanted* to nurse! Once, when we were living in the bush and away from other people, I was especially full of tears. I cried all the time. That was when my father said he was going to beat me to death; I was too full of tears and too full of crying. He had a big branch in his hand when he grabbed me, but he didn't hit me; he was only trying to frighten me. I cried out, "Mommy, come help me! Mommy! Come! Help me!" When my mother came, she said, "No, Gau, you are a man. If you hit Nisa you will put sickness into her and she will become very sick. Now, leave her alone. I'll hit her if it's necessary. My arm doesn't have the power to make her sick; your arm, a man's arm, does."

When I finally stopped crying, my throat was full of pain. All the tears had hurt my throat. Another time, my father took me and left me alone in the bush. We had left one village and were moving to another and had stopped along the way to sleep. As soon as night sat, I started to cry. I cried and cried and cried. My father hit me, but I kept crying. I probably would have cried the whole night, but finally, he got up and said, "I'm taking you and leaving you out in the bush for the hyenas to kill. What kind of child are you? If you nurse your sibling's milk, you'll die!" He picked me

> up, carried me away from camp and set me down in the bush. He shouted, "Hyenas! There's meat over here. . . . Hyenas! Come and take this meat!" Then he turned and started to walk back to the village.
>
> After he left, I was so afraid! I started to run and, crying, I ran past him. Still crying, I ran back to my mother and lay down beside her. When my father came back, he said, "Today, I'm really going to make you shit! You can see your mother's stomach is huge, yet you still want to nurse." I started to cry again and cried and cried; then I was quiet again and lay down. My father said, "Good, lie there quietly. Tomorrow, I'll kill a guinea fowl for you to eat."

Nisa's father hoped this would do the trick, but it didn't: "The next day, he went hunting and killed a guinea fowl. When he came back, he cooked it for me and I ate and ate and ate. But when I was finished, I said I wanted to take my mother's nipple again. My father grabbed a strap and started to hit me, 'Nisa, have you no sense? Can't you understand? Leave your mother's chest alone!' And I began to cry again."[42]

What happened soon after is reminiscent of Inuit use of hypotheticals to ground children in thinking about major and stressful moral dilemmas, except that Nisa's dilemma was far from being merely hypothetical. Her mother was contemplating infanticide right after the younger sibling's birth, and rather than shielding little Nisa, she involved her very directly:

> Mother's stomach grew very large. The first labor pains came at night and stayed with her until dawn. That morning, everyone went gathering. Mother and I stayed behind. We sat together for a while, then I went and played with the other children. Later, I came back and ate the nuts she had cracked for me. She got up and started to get ready. I said, "Mommy, let's go to the water well, I'm thirsty." She said, "Uhn, uhn, I'm going to gather some

> mongongo nuts." I told the children that I was going and we left; there were no other adults around.
>
> We walked a short way, then she sat down by the base of a large nehn tree, leaned back against it, and little Kumsa was born. At first, I just stood there; then I sat down and watched. I thought, "Is that the way it's done? You just sit like that and that's where the baby comes out? Am I also like that?" Did I have any understanding of things?"

Many hunter-gatherers practice infanticide, and the way they deal with this act morally is to dispose of the infant immediately, before they consider it a human being. In this respect, Nisa was in for a frightening surprise:

> After he was born, he lay there, crying. I greeted him, "Ho, ho, my baby brother! Ho, ho, I have a little brother! Some day we'll play together." But my mother said, "What do you think this thing is? Why are you talking to it like that? Now, get up and go back to the village and bring me my digging stick." I said, "What are you going to dig?" She said, "A hole. I'm going to dig a hole so I can bury the baby. Then you, Nisa, will be able to nurse again." I refused. "My baby brother? My little brother? Mommy, he's my brother! Pick him up and carry him back to the village. I don't want to nurse!" Then I said, "I'll tell Daddy when he comes home!" She said, "You won't tell him. Now, run back and bring me my digging stick. I'll bury him so you can nurse again. You're much too thin." I didn't want to go and started to cry. I sat there, my tears falling, crying and crying. But she told me to go, saying she wanted my bones to be strong. So, I left and went back to the village, crying as I walked.

Nisa was involved in a real-life dilemma, from which she was not shielded, and she was now bonded with her younger sibling:

I was still crying when I arrived. I went to the hut and got her digging stick. My mother's younger sister had just arrived home from the nut groves. She put the mongongo nuts she had gathered into a pile near her hut and sat down. Then she began roasting them. When she saw me, she said, "Nisa, what's wrong? Where's your mother?" I said, "By the nehn tree way out there. That's where we went together and where she just now gave birth to a baby. She told me to come back and get her digging stick so she could . . . bury him! This is terrible!" and I started to cry again. Then I added, "When I greeted him and called him 'my little brother' she told me not to. What she wants to do is bad. . . . That's why I'm crying. Now I have to bring this digging stick to her!"

My mother's sister said, "Oooo . . . people! This Chuko, she's certainly a bad one to be talking like that. And she's out there alone with the baby! No matter what it is—a boy or a girl—she should keep it." I said, "Yes, he's a little boy with a little penis just resting there at the bottom of his stomach." She said, "Mother! Let's go! Let's go and talk to her. When I get there I'll cut his umbilical cord and carry him back."

I left the digging stick behind and we ran to where my mother was still sitting, waiting for me. Perhaps she had already changed her mind, because, when we got there, she said, "Nisa, because you were crying like that, I'll keep the baby and carry him back with me." My aunt went over to Kumsa lying beside my mother and said, "Chuko, were you trying to split your face into pieces? You can see what a big boy you gave birth to, yet you wanted Nisa to bring back your digging stick? You wanted to bury this great big baby? Your own father worked to feed you and keep you alive. This child's father would surely have killed you if you had buried his little boy. You must have no sense, wanting to kill such a nice big baby."

My aunt cut his umbilical cord, wiped him off, put him into her kaross, and carried him back to the village. Mother soon got up and followed, shamed by her sister's talk. Finally she said, "Can't

> you understand? Nisa is still a little child. My heart's not happy that she hasn't any milk to drink. Her body is weak. I want her bones to grow strong." But my aunt said, "When Gau hears about this, he'll beat you. A grown woman with one child following after another so nicely, doesn't behave like this." When we arrived back in the village, my mother took the baby and lay down.

Bushmen and other foragers may not use hypothetical moral dilemmas systematically to socialize small children as the Inuit do. But children who live in such intimate groups, where so little remains private, are still likely to be emotionally involved in serious real-life moral dilemmas that involve them directly or indirectly. I suggest that this provides a universal means of helping children to internalize their group's values and apply these values to everyday situations, and that all the Inuit teasing with "hypotheticals" is merely a brilliant—if stressful—way of manipulating and intensifying this natural learning process.

After this, Nisa's competitive desire to continue nursing was dealt with in terms of rules she was supposed to follow, but her ambivalence seems to have continued unabated:

> After Kumsa was born, I sometimes just played by myself. I'd take the big kaross and lie down in it. I'd think, "Oh, I'm a child playing all alone. Where could I possibly go by myself?" Then I'd sit up and say, "Mommy, take my little brother from your kaross and let me play with him." But whenever she did, I hit him and made him cry. Even though he was still a little baby, I hit him. Then my mother would say, "You still want to nurse, but I won't let you. When Kumsa wants to, I'll let him. But whenever you want to, I'll cover my breasts with my hand and you'll feel ashamed."

Here we have an explicit mentioning of shame, which means that the mother is linking rules of conduct with moral feelings. This obviously contributes to Nisa's moral socialization. It seems likely that with hunter-gatherers more generally, a small child's rule internalization may often first begin just with a sensitive but highly intuitive awareness of parental approval versus disapproval of specific acts in self and others. Later the process becomes engaged with manipulative verbal instructions as to how to behave and references to impropriety and shamefulness.

We must not allow Nisa's complaint-oriented autobiographical style to obscure the fact that just as Briggs suggests, Nisa eventually was able to face this moral dilemma of sibling rivalry and resolve it in the light of internalized values that favored being generous within the family. Indeed, as adults she and her younger brother, Kumsa, were quite close. But the childish dilemma she faced was real, not hypothetical, and it seems to have enabled her to work through a problem arising from her own ambivalence and behave eventually in a socially acceptable way. This took place in a society in which both nepotistic and altruistic generosity were openly and pointedly praised, and as Nisa internalized these values, this likely helped her to get past what seems to have been a nearly traumatic experience.

I believe that all hunter-gatherer societies offer such learning experiences, not only in the real-life situations children are involved with, but also in those they merely observe. What the Inuit whom Briggs studied in Cumberland Sound have done is to not leave this up to chance. And the practice would appear to be widespread in the Arctic. Children are systematically exposed to life's typical stressful moral dilemmas, and often hypothetically, as a training ground that helps to turn them into adults who have internalized the values of their groups.

THE COEVOLUTION OF GENES AND CULTURE

The habit of discussing innate dispositions and culture as separate entities goes back to Darwin and even beyond, and this can be useful analytically. However, as anthropologist William Durham[43] has nicely exemplified, behaviors ranging all the way from lactose tolerance to incest prohibitions[44] can be better understood as the product of both in combination, and his highly detailed work on incest will come to the fore in the next chapter.

The egalitarian syndrome that contextualizes the cultural learning of LPA foragers has been in place long enough for an ever-improving fit to develop between useful cultural practices, which are shaped by moral communities, and the useful genetic predispositions that make such practices very easy to learn. This can be readily assumed—even though for humans scientists can tell us very little about behavior genes per se. Thus, whenever we find a behavior that is universal among the fifty foraging societies I have studied, we can appropriately ask if it is likely to have some substantial (if less than wholly "determinative") genetic preparation.[45]

One rule of thumb might be that if a behavior is universal, and if it also goes back to antecedents reconstructed for Ancestral *Pan,* and if its adaptive benefits to individuals can be explained logically, it's likely to be rather well prepared by genes. This is true of early language acquisition, and likewise there are the predictable stages of moral development we have seen for children. Ancient predispositions at least include a child's using its mother as a behavioral model, along with a primitive capacity for self-recognition, innately based dominance and submission tendencies, and also a strong resentment of being dominated. For instance, as children today on playgrounds we readily learn to not only form pecking orders but also actively join in subordinate coalitions to work against the dominance of powerful individuals, and sometimes we pick on newcomers. I remember this well.

Learning to follow rules is something else that we readily learn when our development reaches a certain point, and this, too, was true of Ancestral *Pan.* Then there are the aforementioned human stages of moral readiness that prepare infants and small children to color with embarrassment and blush with shame and later help them to learn complex rules of conduct through playing children's games. Mel Konner believes that very significant rule learning comes through these children's games, as studied by Piaget,[46] and such games appear to be universal. In fact, young apes I've watched spend a lot of their time play-fighting, which teaches them how to express and control their aggressions. And their mothers impose "rules" on them such as not playing too roughly.

The games of human children all have rules, even though the specific games and rules vary greatly from culture to culture. And from our own earlier experiences on the playground, we know that as games are learned, children make newcomers aware of rules and deliberately instruct novices or those who inadvertently break them.

I remember as a lightly built first-grader being taken by my mother (who knew nothing about the game) to the grassy Common in Cambridge, Massachusetts, to join a regular Sunday morning tackle football game that was presided over by one of the boy's fathers. Having never heard of football, and being completely untutored in its rules, I took one look and assumed that the point of the game was to enter a chaotic rough and tumble competition and gain possession of the ball. I proceeded to compete in this way, and at the end of the first play I illegally wrestled the ball away from the downed ball carrier. I was removed from the game diplomatically, but nonetheless I had the distinct sense that I was being treated as a social pariah. And I was, in fact, being excluded, if not aggressively ostracized. The feeling of suddenly discovering that I was a social deviant who flagrantly went against the rules that everyone else was following haunts me to this day, even though a few years later I did learn the rules and came to love the game of football.

In their play, children learn a great deal about things like how to apply their dominance and submission tendencies and how to form political alliances. Sometimes they behave this way to enhance their own power, but sometimes they form coalitions to punish deviance in the form of cheating or bullying. Thus, games turn children's groups into minimoral communities that provide learning experiences in dealing with adult rules of conduct. The internalization of rules appears not to end with childhood. Adults' active and universal verbal amplification of altruism, which we have discussed at some length, works off of very much the same innate tendencies. It reinforces earlier teachings in favor of being generous, and participation in the punishment of behaviors like stealing, cheating, and bullying has a reinforcing effect.

When Cephu was being confronted and went to sit down as usual by the fire, the Mbuti hunters who faced him down knew exactly what they were doing when they denied this cheater his customary place. They were acting as contributing members of a punishing group, and their feelings about having been cheated were especially strong because the rules for fair meat-sharing were embedded so deeply in people's psyches. Cephu's meat-cheating went against a well-learned and strongly believed in sharing ethic that was held by all band members, and the result was systematic collective shaming and a real threat of banishment.

In this type of experience it is quite difficult to separate out the genetic preparations and the cultural input, for the two are closely intertwined. But because the cultural glove fits the genetic hand so well, social selection was able to have consistent effects on both group social life and gene pools over many generations and, in fact, over scores of millennia.

Work of the Moral Majority 9

GOSSIP'S TYRANNY

Morality's a group affair, but often individuals take the lead in exerting social pressure, and, when that doesn't work, in initiating more assertive sanctions against deviants. Critical to this process is talking, and gossip's a special kind of talking. We're going to talk a lot about gossip here, for it's because of language and gossiping that foraging bands are so likely to exhibit the universal, well-developed, and highly negative expressions of "public opinion" that are mentioned so regularly by anthropologists who live with small groups of nonliterates.

Gossiping tends to have a bad name with ourselves and hunter-gatherers alike, even though public opinion can also be focused on the positive qualities of "gossipees" as well as their faults. Most of this "talking" is in fact negative in its thrust—as is demonstrated anthropologically by John Haviland's linguistic study of colorfully obscene gossiping in a Mayan peasant village.[1] Whether they like it or not, this shapes people's social reputations in Yucatan—and they'd better watch out, for gossip functions as a court of public opinion. It's a special kind of court, however: the defendants don't get to face the charges against them—and often there's simply no way to defend themselves.

The same predominant negativity is true of hunting bands. The critical court of public opinion isn't interested in holding balanced and fair hearings but rather in knowing about what people will try to hide. Anthropologist Polly Wiessner has made a detailed study of what Kalahari Bushmen talk about in the context of individual culpability, as a possible prelude to group social pressure or punishment.[2] It appears that forager gossip is similar to Haviland's peasant gossip and to the Serbian tribal gossip I listened to daily with such great interest—and, I admit it, great enjoyment—in Montenegro.

The !Kung people that Wiessner has studied for over three decades gossip intensively when trouble is shaping up and collective action may have to be taken. Her uniquely long-term data show what is talked about, and the leading social problem has, of course, been twenty-nine cases she counted of "big-shot" behavior—a pattern that creates a major social challenge in any egalitarian society. As we've seen in Chapter 4, when excessively dominant behavior seriously threatens the personal autonomy of others, this can result in capital punishment. And even moderate signs of big-shot behavior are certain to be of keen interest in egalitarian hunter-gatherer gossiping circles everywhere, for such behavior is best headed off at the pass.

Other !Kung problems that stimulated group "talking" and then possible collective action were, in order of incidence, (1) patterns of stinginess, greed, or laziness; (2) an individual acting as a troublemaker; (3) political or land-use disputes; and (4) someone being reclusively antisocial. Other problems included inappropriate sexual behavior. The two most frequent—big-shot behavior and patterns of stinginess, greed or laziness—can both be seen as free-riding attempts in that generous good citizens will be taken advantage of. Thus, Wiessner's study provides excellent corroboration for what I've said earlier about the LPA focus on reduction of free-riding behavior.

It's by adding up information that social deviants are identified and people can unite to cope with them. Without safe, private gossiping, free-rider suppression would not be likely to work very effectively in the case of scary bullies, because only a united group is a confident and safe group, and such political unity comes out of finding a consensus. Furthermore, a constantly communicating group is one that can quickly figure out diagnostically who is a thief or a lying cheater. Thus, effective punitive social selection is possible because a band's better citizens can communicate in private, in a situation of trust.

Polly Wiessner has been visiting the !Kung Bushmen now for three decades, so on average this means that about once a year there's a serious incident in which big-shot behavior is sufficiently worrisome that talking might lead to collective action. With respect to free-rider suppression, this means two things. First, the bullying free riders can be confronted by the group; this gives them a chance to reform and stop the behavior before it brings on dire consequences. And second, others harboring the same antisocial tendencies can learn by example and simply desist from the pushy or arrogant behaviors that come all too naturally to certain male hunters. Having an evolutionary conscience makes it easy to make the necessary calculations.

In this way groups often can prevent individuals who are unusually disposed to take a bullying free ride from carrying things to the point that their reproductive losses will become severe owing to capital punishment or other decisive and injurious sanctioning. However, even if collective action by the group leads to reform, the reformed deviants will have paid some costs because they've been temporarily excluded from cooperation, and there may be reputational costs to be paid in the future. The potential bullying free riders whose genes suffer no adverse consequences are those whose evolutionary consciences are so effective that they can efficiently inhibit themselves in the first place, keep people from feeling threatened, and maintain a decent reputation.

As we've seen, Jean Briggs's intimidating adoptive Inuit father, Inuttiaq, had some decided difficulties in this area, but he was able to keep his potentially self-aggrandizing volatility in check. To stay out of political trouble, such an unusually assertive person must be aware that the rest of the group will be watching him carefully and will be privately comparing notes to make sure they can trust him to keep his aggressive tendencies under control.

Thus, there are a variety of outcomes for individuals disposed to bullying behavior. Some must be killed, and many others must be ridiculed or admonished and thereby reformed. But many have consciences that are up to the task of staying in line and can do this so well that group sanctions never come into play. Awareness of gossip and fear of it are part of this picture, for other people's talking privately can have significant social consequences for deviants, be they bullies, cheaters, or whatever. And because everyone in a band is privy to gossip, all band members appreciate their own personal vulnerability to this type of private discourse.

Gossiping in private is not necessarily safe. The gossipers must still be careful about what they say, for if negative pronouncements are repeated, or overheard, this can spell real trouble both for themselves and for the entire group. (It was such a breach of confidentiality that triggered the infamous duel between Aaron Burr and Alexander Hamilton.) Gossip can seriously damage a person's reputation—and everyone knows that bad news, once started, travels rapidly. In my survey of hunter-gatherer societies, I actually found one case in the Arctic of an inveterate gossip's being killed by her group, presumably because her malicious talk stirred serious and unnecessary conflicts.[3]

I could write a book about gossip. Any cultural anthropologist knows that until you begin to access the gossip networks in the group you're studying, you'll have little chance of understanding what's really going on socially. Eliciting gossip was one aspect of the mid-1960s fieldwork with Montenegrin Serbs that both my wife and I enjoyed immensely once we were plugged in, not only because we were finally getting the "real story" about tribal affairs and people's moral reputations, but also because reaching the point of having such intimate and trusting conversations made us feel we were gaining some real acceptance from the people we had chosen to work with. We were making friends, and friends who trust one another just naturally gossip.

In the tightly networked settlement where we were living, it was all too apparent that the preponderance of this "talking" was focused negatively on individuals suspected of deviant behavior.[4] It also seemed, shades of Durkheim, that virtually all the settlement's inhabitants felt they had *something* to hide and feared the social effect of others talking about them and jumping to wrong—or right—conclusions. This fear was realistic, for all this private talking involved zealous social "detective work" that could bring down reputations. As a result, people not only loved to gossip because of the insights they gained, but they also fiercely resented the fact that at the same time others surely were talking privately about them.

In the tribe we studied, one man whose proper name I shall not divulge was known as a *prepritzalitza,* which literally means "one who tells things over and over again," or perhaps better, a "serial gossiper." This man of fifty deviated from the local standard not because he so obviously enjoyed prying into other people's affairs and spreading bad news, but because in doing so, he didn't confine his gossip to small, careful, private networks of trusted and discreet close friends. Rather, at a sizable evening social gathering in front of a whole roomful of people, he'd start broadcasting information about individuals who were absent just as though he were in a tête-à-tête situation. There was never talk of taking action against him, but he was cordially disliked, and he himself was talked about for being a reckless gossip who *openly* endangered the reputations of others. Even though this lifelong behavior had seriously damaged his reputation, "Svetozar" was never actually confronted, probably because his loose talk wasn't seriously disruptive to the settlement he lived in. Indeed, it seemed to be driven more by naïveté about how things worked socially than by any extraordinary degree of malice.

Normal gossiping serves as an agency of indirect social control. And aside from scaring many people sufficiently to deter them from discoverable acts of deviance, gossiping has other social benefits, as well. Over time, highly useful social information can be disseminated widely until a group consensus forms. Thus, for instance, a case can be built gradually against a main suspected thief where several other candidates were also under suspicion initially. I watched this happen in the remote Serbian tribal settlement we studied, and the process went on for months, all but obsessively.

Let's return to bands. If a serious problem arises, it's the combination of the group's shared rules and a consensus about the actual facts of a particular case that enables a band to come down unanimously on a deviant like Cephu—or at least to come down on him as a largely united moral majority while the deviant's close kinsmen may choose to stand to one side. This singularity of purpose is important because if a consensus is not built before action is taken, what might otherwise have been an instance of efficient group sanctioning can turn into sheer factional conflict, with *both* sides claiming moral rectitude. And this has the obvious effect of damaging the social fabric of the group, whereas putting an end to a seriously deviant pattern of behavior, doing so in the name of the group, will greatly improve that fabric.

Being tied in to intimate gossip networks surely has been individually adaptive ever since humans with some kind of symbolic language began to systematically identify and deal with intimidating or hard-to-detect deviants by privately comparing notes, thereby making them easier to avoid and more likely to pay for their depredations. Furthermore, when children overhear their parents gossiping, this provides their developing consciences with accurate information about values, rules, and proper behavior. This means that language has had an important role in moral evolution and values internalization, if perhaps not an absolutely necessary one: in resisting alpha domination, a nonmoral Ancestral *Pan* at least had the potential to find a political consensus through use of body language, nonsymbolic vocalizing, and astute reading of social contexts.

This capacity was evident in the group-sanctioning episode Frans de Waal interpreted for us when he told how the protesting voices of the female chimpanzees in his captive group reached unanimity and the alpha male knew that a physical attack was coming if he continued his bullying behavior. This was not close to being gossip, for the emotional outcry that resulted from "opinion sharing" was totally public and it lacked any symbolic expression—but the overall social context did give the communication obvious meaning in terms of a shared social diagnosis and shared hostility, with a successful past history of rebellious action against bullies providing part of the shared meaning. I believe that these collectivized angry communication functions in great apes at least have some basic similarities to what humans do in gossiping, where a *moral* consensus, based on explicit shared values and rules, leads similarly to social control.

Had earlier humans been totally unable to communicate privately through symbols, it's unlikely that anything very similar to moral communities as we know them today could have evolved. The fact that a solid public opinion can be quietly formed without personal risk or conflict and through highly specific symbols provides a formidable social tool, especially against dangerous bullies, and when it's time for action, the findings can be used surgically.

I've mentioned, in passing, gossip's staying power as society modernizes. This appears to involve more than the purely habitual perseveration of a hunter-gatherer cultural habit. Indeed, it's likely, after at least 45,000 years of individuals gaining fitness by whispering back and forth about other people's behavior behind their backs, that discreet and intimate socially evaluative "talking" is part of our evolved capacity to behave morally—just as the internalization of rules surely is. Thus, gossiping serves us today as it served our ancestors in the past.

I'm not suggesting that there's anything like a single gene for gossiping or that there's a specific brain area—a "module," as it's sometimes called—that's dedicated to informative backbiting. What I do propose is that such behavior is likely to be both ancient and innately fairly well prepared, precisely because individuals whose gossip networks are superior can gain a certain edge—both in staying out of serious trouble themselves and in better learning which social predators to watch out for. Gossiping creates fascinating information that is useful in everyday life and also helps fitness. That's why it's so prevalent and persistent. That's also why millions of modern females (and *many* more males than you might think) watch soap operas—without quite understanding why they lock in so readily.

KEEPING DEVIANTS AT A DISTANCE

The need to share large game is one reason that people cluster together to live in bands. But there are others. Humans are naturally sociable, so these nomads also congregate simply because they enjoy the company; that way there's also better protection against predators. Furthermore, bands link people with resources: when two bands recognize each other's collective claims to specific, highly familiar natural resources, this aids subsistence efficiency on both sides. All of these factors militate for a minimum band size of twenty to thirty people and for a certain unity of purpose.

Because bands cooperate so strikingly, we tend to think of their members living in great harmony. However, even though in general these several dozen people will camp together and live in close physical proximity, and even though they have a shared subsistence fate as large-game hunters, some of a band's families will choose to live closer to certain families than to others. I'm referring to physical space, but they also have a variety of ways of "distancing" themselves *socially*.

Social distancing theory began with the study of modern ethnic groups and their views of one another as "outsiders," but I will be expanding this theory to include insiders treating other insiders as outsiders.[5] With respect to factions within groups, Cephu contentiously referred to his little cluster of intimates as being separate from the rest of the Pygmy band, and in fact they built their shelters accordingly. The Kalahari Bushmen orient the openings of their huts according to feelings of social closeness or distance,[6] and Briggs[7] tells of an Utku family that was spatially distanced from the rest of the small band largely because one family member, a female named Niqi, had a seriously problematic personality. This included a free-riding lack of generosity, poor social skills, and a general emotional volatility that was distasteful to other Utku, and although Niqi's whole family merely camped at a distance, Niqi herself was subject to some moderate ostracism.

Active social distancing by entire bands is aimed at individuals whose deviant behavior annoys, outrages, or threatens other group members, and most of the predictable reactions have already been described. It all begins with a social aloofness that reduces normal communication, as in curtailed everyday greetings by group members. Such reticence pertains also to conflicts just between pairs of individuals, and one way to keep a dyadic conflict from escalating is simply for the two to stop speaking bilaterally. When everyone in the band starts to restrict greetings and verbal intercourse, this curtails the deviant's opportunities to enter into cooperation, while the social deprivation will be a major source of stress—unless the person chooses to move to a different band, which may or may not be feasible.

Although ostracism is fairly frequent and surely is universal among LPA foragers, the details are seldom reported in any detail. We are fortunate, therefore, to have the account from Jean Briggs of her own ostracism in an isolated Inuit camp of less than two dozen people. It's hard to say just where it all began, for basically the Utku were upset with Jean's emotional style long before the defining incident took place. For instance, if the igloo's roof melted and a hunk of slush fell into her typewriter, she was prone to throw something (in one instance, a knife) simply to express her anger—a behavior that was frightening to the Utku because in their minds killing someone came next.[8] In Jean's New England culture of origin, this would, of course, be a mere fit of pique.

Here, greatly abbreviated, is the account I wrote in *Hierarchy in the Forest* of Jean's unfortunate but illuminating ordeal. Inuttiaq, the man who has the problems with an overly assertive personality, is the band's informal headman.

> Several sportsmen who had flown in by seaplane were borrowing the Utku's two rather fragile, irreplaceable canoes for fishing. The

Utku approach to such exploitation, which they resented in spite of some trading with the whites, was to acquiesce to every request. Having been told in private about their resentment, Briggs made the mistake of actively intervening after the whites ruined one of the Utku canoes and still wanted to use the last one. She explained heatedly that the Utku depended on the canoes, and the guide replied that if the canoe's owner did not choose to lend the boat they would do without it. Inuttiaq, put on the spot by Briggs, agreed to let the sportsmen use the last boat for fishing. Briggs could not hide her anger at both parties. She strode away from the scene to weep in her tent, unaware that her behavior had been the last straw. . . .

After the incident people continued to behave cordially and generously toward the turbulent visitor when she approached them—but she noticed that they were coming to visit her far less often, and stayed only briefly. Before long they began in subtle ways to discourage her own visits. It became evident that she was being ostracized like Niqi—but during the period of estrangement it also emerged that the Utku were willing to let her rehabilitate herself. Indeed, whenever she was able to control her feelings and avoid emotionalized negativity for a time, they began to respond positively. The problem was that Briggs could not maintain the flawless equanimity that the situation seemed to demand. She was being distanced, it hurt, and when she broke down and showed her frustration, she was distanced even more.

The kind of overt hostility that had triggered the ostracism was scrupulously avoided by the Utku in applying their sanction. Briggs was basically alone in her tent with very few callers, and even their visits became perfunctory compared to her earlier rich social life. People were not impolite, they simply set up a great deal of social distance. By good luck, after months of this treatment a third party provided the Utku with some insight into why Briggs had tried to deny the canoe to the whites. Once

> they understood that her anger had been on their behalf and against unscrupulous men who cared nothing for their welfare, they eventually relaxed the social barriers they had so carefully erected and maintained.[9]

Jean was not *shunned* in the sense of an overtly hostile cutoff of all social intercourse such as takes place in Lancaster County, Pennsylvania, with the Amish.[10] But it's difficult to imagine being thousands of miles from your own culture and being socially rejected in this low-key but hostile way. People who engage in social ostracism know they're inflicting emotional pain, and they tend to do so manipulatively—for generally they're willing to permit rehabilitation. That's the beauty of ostracism and also of all-out shunning, for neither is necessarily forever. It's banishment that is forever.

In *Never in Anger* Jean interprets this act of social distancing in terms of the Utku fear of potentially violent emotions, which they themselves are so careful to control. However, there may have been another element that I might add. In going against the group's decision, she may have been perceived as a person who was trying to make a decision individually for an entire egalitarian band, a consensus group that allowed none of its own to do this. Jean definitely was far too free in letting out her anger, but in Utku eyes she also may have been acting something like a big shot when she unilaterally presumed to speak for the entire group and thereby went against its wishes. The price she paid was that she was politely but firmly denied normal social contact for a matter of months.

Social scientists usually categorize such distancing in escalating terms of ostracism, shunning, temporary group ejection, and permanent banishment, but in fact capital punishment can be seen as the truly ultimate means of social distancing.[11] Of course, taking a life unnecessarily within the group is a strong moral prohibition, but group-ordained capital punishment isn't. And even though the ethnographic coverage is so incomplete, I believe that such punishment is all but certain to be potentially present in any band society, as a desperate final resort. Thus, social distancing usually is manipulative in the direction of reform, but as we've seen it also can become ultimate.

ACTS OF "SOCIAL MONSTROSITY"

Unlike flagrant meat-cheating, some morally deviant acts seem not to *directly* threaten the interests of all group members, yet they, too, may elicit a collective response of lethal violence because they are, to use a nontechnical term, socially monstrous. Close incest is often viewed that way, although cultures vary considerably in how they define this social crime. This variation may be driven in part by the degree to which in the society's past inbreeding's consequences were consciously recognized.

Evolutionary anthropologist William Durham studied incest using a sample of five dozen societies, most of them tribal but with a few foragers, and in nearly half of this sample there was evidence to suggest that somehow incest taboos had been created because at some point in the past birth anomalies were noticed and associated with close inbreeding.[12] In addition, a general human capacity for disgust could be hooking up with these strong moralistic reactions.[13] However, I certainly would hesitate to suggest that there could be a very *specific* innate horror of, say, mother-son incest, even though this prohibition appears to be the strongest.

Sibling incest usually elicits abhorrence, but there are rare documented instances of socially approved brother-sister marriage among royalty (and others) in early civilizations. Especially in tribal societies, often marrying a true first cousin is heartily approved of, sought after, and strongly institutionalized. Thus, if we look at all world societies, the application of close-incest taboos is somewhat flexible even though mating between parents and children is always "against the law."

To complicate matters, there's the Westermarck Effect,[14] which predicts that when children are raised in siblinglike relationships, whether they are very closely related or not, they will develop a mutual sexual inhibition that kills attraction and therefore radically reduces their breeding possibilities. This effect has been measured statistically,[15] and similar effects are seen in promiscuously breeding great apes with respect to socially familiar relationships between mothers and sons or between siblings, both of which are naturally inhibited.[16] However, if such an evolved inhibition is at work in humans, it doesn't seem to be extremely strong with respect to the father-daughter incest relationship, for the rates in modern American society would seem to be noteworthy, with Appalachia as everyone's favorite example.[17] On the other hand, modern society provides far more privacy than does band life.

Durham found that active sanctions like ostracism, physical attacks, and even execution were measures taken against close incest as locally defined. Less often supernatural sanctions were an agency of social control. For example, there might be fear of being born with birth anomalies inflicted by imaginary entities. In some cultures, because these scary imaginary entities are thought to punish the entire group for the incestuous sins of individuals, band members will feel incest to be a serious threat to everybody and therefore punish it very severely.

Whatever the puzzles about incest, there appear to be two bases for punishment of monstrous acts by outraged moral majorities. One is that the deviance creates a clear and present danger to every other member of the group, either realistically or supernaturally. The other is that somehow the behavior is found to be so far beyond the pale that it simply becomes morally repugnant. In that sense, close incest and serious meat-cheating may be fairly similar, for they can fulfill both of these criteria.

KILLING UNBALANCED DEVIANTS

Group opinion can come into play actively and lethally in the absence of *moral* malfeasance. There are rare cases where individuals who are patently unable to understand the damage their behavior is doing to others must be dealt with by their groups, and usually this is because group members feel seriously threatened for their lives—as with a psychotic person whose violent nature can turn to homicide. There are both Eskimo and Bushman examples. Here's an Inuit case history from Balikci:

> Shortly after, Arnaktark must have returned to his igloo and that same night he stabbed his wife Kakortingnerk in her stomach. She fled on foot with her child on her shoulders, and after arriving at the main camp she told what had happened.
>
> They started to fear that he might stab again at someone they loved, and they discussed what should be done. The discussion was held among family, and it was felt that Arnaktark, because he had become a danger to them, should be killed. Kokonwatsiark [a brother] said that he would carry out the verdict himself and the others agreed. Old father Aolajut was not supposed to do it, because Arnaktark was his own son; but if Kokonwatsiark for some reason would not have done it, the next oldest, Abloserdjuark, would have offered himself to do it. After the decision was taken, Kokonwatsiark notified the non-relatives, because they also were afraid. All agreed that there was no alternative.
>
> Then the entire camp broke up: Aolajut, Kokonwatsiark, Abloserdjuark, Nerlongajok and Igiukrak traveled to Arnaktark's igloo, and Krimitsiark led the others and the women and children along another route to the new camp at the coast. Upon arrival at Arnaktark's place, the latter was standing outside, and Kokonwatsiark said to him: "Because you do not know very well any more (have lost control of your mind), I am going to 'have' you." Then he aimed at his heart and shot him through the chest. Then they moved on to join the others at the coast. His grave is yonder, towards the end of Willerstedt Lake.[18]

This provides a sad example of a Netsilik delegated killing in that there was group agreement and a kinsman was the agent. However, the execution was expedient, rather than moral, because the deviant didn't understand his own actions. There's a !Kung Bushman case history, also with exceptional ethnographic detail, that I shall bring in later in the chapter. However, in that case it's a bit less clear as to whether the killer was considered to be "psychotic," and to say that the manner of execution was far less orderly is an understatement.

Given enough time, any band society is likely to experience a problem with a homicide-prone unbalanced individual. And predictably band members will have to solve the problem by means of execution even though all of these societies are strongly against killing within the group. As with infanticide or the frequent killing of twins, this is not capital punishment because moral malfeasance is not the issue. But there seems to be a similarity to capital punishment in that the Inuit act of execution was delegated to a family member, and in that a serious threat to everyone in the band begot community-sponsored action.

When a rational but incorrigible, selfish bully is taken out by his peers, the social and political dynamics are the same—but a moral element is added. In Table I, we saw that such executions seemed to be far more numerous than the killing of psychotics, so they, too, may be assumed to be universal in LPA societies. Give any of these groups a few hundred years to work with, and a reckless male, intent on domination, will all but certainly turn up and have to be coped with. And sometimes killing him is the only way out.

PEACEMAKING AS A DIFFICULT MISSION

Scientists have often tended to view hunter-gatherer levels of violence as a window into our genetic nature, and of course the less violent our favorite forager example happens to be, the more we're in luck with respect to thinking we must have a nice, as opposed to a nasty, human nature. Earlier studies of the Bushmen culminated in a book called *The Harmless People,* and even though the author, Elizabeth Marshall Thomas, was aware that the contentious !Kung were always worried about having poisoned arrows around, she wrote that homicides were rare[19]—even though back then they'd been killing one another at the same rate as New Yorkers or Los Angelenos.

Thomas did accurately characterize the !Kung as people who are sensitive to group opinion and eager to resolve conflicts; indeed, when it came to lower levels of conflict, these people were likely to hold an all-night trance dance, which meant that the entire band would dance and sing together and restore the social harmony they valued. As we've seen, however, in the matter of serious conflicts and murder rates, Thomas's information, like that of many others who practice the scientific art of ethnography, was seriously incomplete.[20]

It was only later, after Richard Lee finally persuaded a few !Kung informants to talk to him openly about homicide, that the facts of Bushman killing came spilling out, and it became apparent that these people, who are deeply against conflict in their values, have quite high homicide rates because they lose their tempers easily and are experts at killing sizable mammals. The Marshall expeditions began just as Bushman conflict was beginning to be greatly inhibited by state control, but later, by turning to ethnohistory, Lee was able to write about these guarded but well-remembered earlier conflicts as a prominent part of the Bushmen's traditional social fabric.[21]

As with all forager societies, there's a local cultural style for expressing personal differences among the !Kung. This begins with minor but potentially serious arguments that are confined to hostile joking, with merriment relieving the tensions. At the next level, angrier disputants drop the pretense of humor and the potential for violent conflict increases as arguments become more heated. Next comes *za,* or sexual-insult exchange, which means that violence is precariously near. For males the ultimate sexual insult is "May death pull back your foreskin," whereas for females it is "Death on your vagina."[22] Because physical exposure of the body parts is shameful for the !Kung, such "verbal exposure" involves implications of shame, and such insults can lead in contrary directions. One, perhaps rather surprisingly, is suicide—presumably with the anger turned inward. More expectably, the other is retaliatory, violent, and potentially very harmful to the person who did the taunting.

Because the males are always packing poisoned arrows, a sudden and potentially lethal assault can be in the offing whenever a verbal conflict reaches the *za* level. When this happens, the band tends to quickly split up because everyone knows that a homicide is coming next, and as with the Netsilik, and as with all hunter-gatherers, apparently,[23] further killing will be likely on a retaliatory basis. Avoidance is a prime method of conflict resolution among all LPA foragers, precisely because their nomadic lifestyle, along with their flexible approach to who can reside in which band, makes this use of physical separation such a practical strategy.[24]

For tribal people who are tied to agricultural land use, it's much more costly to pick up and move, and such sedentary egalitarians are more likely to invest some limited authority in their headmen so that preemptive conflict resolution can become more effective. The !Kung are so zealously egalitarian that this doesn't take place, and when a respected, peace-loving man with no authority tries to step in and stop a *za*-level conflict once it has become physical, Lee tells us that he's likely to be killed himself.[25]

Unlike many LPA foragers, the !Kung seem to squabble about meat all the time, and this might make us wonder about all the claims I've made about sympathetic generosity, "equalized" meat distribution, joy in sharing, and avoidance of conflict. However, this is what happens with the !Kung. A large animal carcass is initially shared out to key people in the band, who then share it further, within large or small family groups, and by then the meat has been transformed by custom from what amounts to group property to personal property.[26] The further sharing that takes place is optional, even though there are expectations and hopes, and along the way there are requests for meat and more requests for meat, and also some accusations of stinginess. But basically the entire band is getting meat, and the overall system is understood and appreciated.

There are very few *serious* disputes over meat.[27] Lee does provide one case history of a homicidal conflict that was caused by adultery but was triggered by a disagreement over meat, but outright forager conflicts about meat are not that frequent, even though the bickering, in certain societies like the !Kung, can be constant. Pat Draper clarifies the Bushmen's sometimes-contentious sharing style as follows:

> Verbal aggression is commonplace among !Kung. In fact, the reason that goods are shared equitably and more or less continuously is that the have-nots are so vociferous in pressing their demands. Are these a people who live in communal harmony, happily sharing among all? Not exactly, but the interpretation of meaning in any culture inevitably founders on these kinds of ambiguities. At one level of analysis, one can show that goods circulate, that there are no inequalities of wealth and that peaceable relations characterize dealings within and between bands. At another level, however, . . . one sees that social action is an ongoing scrimmage—often amicable but sometimes carried on in bitter earnest.[28]

Inuit conflict has been studied almost as intensively as with the Bushmen. By interviewing willing older informants as Lee did, and by making use of reports by early explorers, Geert van den Steenhoven managed to create a reliable account of past homicidal patterns in central Canada.[29] Asen Balikci has drawn on these materials to good effect. Balikci details how the Inuit deal with angry conflicts, and even though their style is quite different from what takes place out on the Kalahari, there are similarly escalating social mechanisms to resolve smoldering or overt conflict.

In a small band, conflict is both stressful and, in its potential, economically costly. In both of these societies, conflicts easily become lethal because neither permits the development of leaders who have enough authority to step in and readily stop a serious quarrel between angry males. (Keep in mind that as killers of large mammals, male hunter-gatherers are armed with very effective hunting weapons.) Balikci writes:

> The Netsilik knew of a number of rather formalized techniques for peacemaking that were positive in the sense that usually they brought conflict into the open and resolved it in a definitive manner. These techniques were fist fights, drum duels, and approved execution.
>
> Any man could challenge another to a fist fight for any reason. Usually they stripped to the waist and the challenger received the first blow. Only one blow was given at a time, directed against temple or shoulder. Opponents stood without guard and took turns, the contest continuing until one of the fighters had had enough and gave up. This seemed to settle the quarrel, for, as one informant put it: "After the fight, it is all over; it was as if they had never fought before."
>
> . . . The song duel was a ritualized means of resolving any grudge two men might hold against each other. The songs were composed secretly and learned by the wives of the opponents. When ready, the whole group assembled in the ceremonial igloo, with a messenger finally inviting the duelists. As was the case with all drum dancing, each wife sang her husband's song in turn, while the latter danced and beat the drum in the middle of the floor, watched by the community. The audience took great interest in the performance, heartily joking and laughing at the drummers' efforts to crush each other by various accusations of incest, bestiality, murder, avarice, adultery, failure at hunting, being henpecked, lack of manly strength, etc. The opponents used all their wits and talent to win the approval of the assembly. . . .
>
> Song duels thus undoubtedly had a cathartic value for the individual opponents, and in this particular sense conflicts became "resolved." Sometimes one or both of the opponents at the end of a song duel continued to feel enmity. When this was the case,

> they often decided to resume fighting, this time with their fists. This definitely settled the matter.[30]

In both cultures homicide is frequent, and such patterns of conflict and conflict resolution may well have been fairly widespread among LPA foragers before contact. In this connection, some of the best Inuit data were collected by early explorers soon after contact, when the people were not yet reticent about such matters. In the rest of the fifty-society database, we've already seen that often ethnographic reports of capital punishment were seriously underrepresented, or sometimes even absent in many of the ethnographies—just as they would have been for the Bushmen had Lee not made his retrospective breakthrough.

My assumption is that prehistorically, lethal conflict would have been a serious potential problem in this type of society, even though local conditions and differing cultural traditions surely impinged on the rates of killing. For instance, the Netsilik had such a high rate of female infanticide (it's a son as hunter who'll support a parent in old age) that some men had difficulty finding wives and therefore were prone to kill another man and take his.[31] This upped the murder rate, but other male killings took place just because personalities clashed and assertive hunters were prone to quarrel. With the Bushmen, women as wives were not so scarce, yet quarrels over courtship or, more often, over adultery provided a regular source of conflict that could result in homicide.[32]

THE COMPLICATED CASE OF MURDER

Murder is conceived of indigenously as a monstrous act. Nevertheless, when an Inuit man slaughters another man in the same band just to take his wife, more often than not his morally outraged band is likely to do nothing about it collectively. Band members know that the victim's close male relatives are likely to take revenge and that usually the killer and his family will quickly get out of town because if he's brazen enough to stay in the band, he may well pay with his life.[33]

Thus, it's revenge slaying by relatives, not capital punishment by the whole band, that he must fear as a first-time offender. This seems to be true of LPA foragers generally, even though our data will never be complete. This retaliatory pattern and the social avoidance it engenders are widespread,[34] and I believe it's the combination of acute loss and grief, combined with deep-seated revenge tendencies,[35] that makes killing for a lost close kinsman (rarely, a kinswoman) so predictable.

Even though murder certainly is not condoned, most initial killings seem not to be felt as a generalized threat by all band members. However, if an individual becomes a *serial* killer, meaning killing two victims or more, then all members of the band are likely to feel threatened on an immediate if probabilistic basis, and they're likely to be galvanized into collective action.

We've seen in Chapter 4 and Chapter 7 that active executions accomplished simultaneously by entire groups are rather rarely reported. They're still more rarely described in any detail, but fortunately we have Richard Lee's specialized study of !Kung Bushman homicides to work with.[36] The !Kung hunt with arrows tipped with poison taken from one of several Kalahari beetles, which can bring down an animal as large as a giraffe after a few days, and which can kill a human much more quickly—unless the wound is shallow and can be quickly lanced to suck out the poison. These foragers are careful to keep their arrows out of the reach of children, lest a heated childish quarrel become lethal, and bystanders will act quickly to try to head off serious quarrels involving two or more adult males because in the heat of the moment one man may grab his bow to shoot another—or may simply stab his adversary lethally with a hand-held poisoned arrow.

Lee's thorough interviews give a degree of ethnographic richness to the statistical patterns we've been describing, and they also help to make clear why delegating a kinsman as executioner is usually preferred over a collective attack by the group—even though either strategy will obviate retaliation by male kin. Keep in mind, here, that we're talking about taking out an intimidating repeat killer who is an expert at killing large mammals as a hunter, who is already experienced at taking human life, and who, because in this case the group poisons its arrows, can still do damage to others while he is dying.

Lee's systematic research encompasses four different regions inhabited by nomadic !Kung from the 1920s through the 1950s, and his case histories include original killings, revenge killings by relatives, group-sponsored executions of repeat killers, and several inadvertent killings of good Samaritans who tried to stop fights that had already become *za*-intense. In this forty-year period, for a region with not a great a number of bands, there were twenty-two killings in all, and in several cases they came in a flurry because chains of retaliation were involved. I emphasize, again, that Lee had the advantage of gaining the trust of informants who had participated in either killings or executions of killers in the period before the government of Botswana began to punish all killing with jail time.

Once the government became very active in this way, in small nomadic camps the !Kung's homicide rates actually dropped out of sight. This supports what I said earlier about hunter-gatherers becoming reticent about their capital punishment practices after contact: knowledge of such organized authority can have a profound effect on these egalitarians, even though they share power equally within their bands and normally make their own decisions about deviants.

The following account is based upon cross-checked interviews with participants who were interviewed by Lee after people became willing to trust him with such information. What we have here is a poorly executed group execution that begins with the use of a poisoned arrow. Lee points out that, even though Bushmen are excellent archers when out hunting, when they get into fights their aiming of poisoned arrows is mediocre at best. In this one uniquely detailed description of what seems to begin as a delegated execution and eventually becomes a fully communal killing, things are so chaotic that it's easy to understand why with hunter-gatherers the usual mode of execution is to efficiently delegate a *kinsman* to quickly kill the deviant by ambush. (As with the exclamation point in front of !Kung, the various diacritical symbols represent several different phonemic "clicks" in the Bushman dialect.)

> The most dramatic account of a collective killing concerns the death of /Twi, a notorious killer who had been responsible for the deaths of two men . . . in the 1940s. A number of people decided that he must be killed. The informant is =Toma, the younger brother of /Twi.
>
> *My brother was killed southwest of N//o!kau (in the /Du/da area). . . . People said that /Twi was one who had killed too many people so they killed him with spears and arrows. He had killed two people already, and on the day he died he stabbed a woman and killed a man.*
>
> *It was Xashe who attacked /Twi first. He ambushed him near the camp and shot a poisoned arrow into his hip. They grappled hand to hand, and /Twi had him down and was reaching for his knife when /Xashe's wife's mother grabbed /Twi from behind and yelled to /Xashe, "Run away! This man will kill everyone!" And /Xashe ran away.*
>
> */Twi pulled the arrow out of his hip and went back to his hut, where he sat down. Then some people gathered and tried to help him by cutting and sucking out poison. /Twi said, "This poison is killing me. I want to piss." But instead of pissing, he deceived the people, grabbed a spear, and flailed out with it, stabbing a woman named //Kushe in the mouth, ripping open her cheek. When //Kushe's husband N!eishi came to her aid, /Twi deceived him too and shot him with a poisoned arrow in the back as he dodged. And N!eishi fell down.*
>
> *Now everyone took cover, and others shot at /Twi, and no one came to his aid because all those people had decided he had to die. But he still chased after some, firing arrows, but he didn't hit any more.*

Then he returned to the village and sat in the middle. The others crept back to the edge of the village and kept under cover. /Twi called out, "Hey are you all still afraid of me? Well I am finished, I have no more breath. Come here and kill me.Do you fear my weapons? Here I am putting them out of reach. I won't touch them. Come kill me."

Then they all fired on him with poisoned arrows till he looked like a porcupine. Then he lay flat. All approached him, men and women, and stabbed his body with spears even after he was dead.

Then he was buried, and everyone split up and went their separate ways because they feared more fights breaking out.

Commentary. This exceptionally graphic account brings out some of the drama of the action. The killer shot so full of arrows that he looked like a porcupine is a remarkable image of the capacity for collective action and collective responsibility in a noncorporate and nonhierarchical society. I interviewed the mother, father, and sister of the dead /Twi and the relatives of his victims. All agreed he was a dangerous man. Possibly he was psychotic.[37]

Interestingly, it was not the man who actively led things off who was killed by /Twi, but a bystander after his bystanding wife was wounded.

If ever there were a flawed "communal" execution, this was it. That's why a delegated close kinsman usually does the job. Close male kin execute oppressive bullies, including maliciously wayward shamans, if such exist, along with the occasional thief, cheater, or nonpsychotic deviant, as seen in previous chapters. In Chapter 7, Table IV showed that out of a sample of ten LPA foragers, six of these societies reported a single group member's stepping forward to fulfill the executioner's role, so we may assume this strategy to be predominant among contemporary hunter-gatherers and also to have been very widespread 45,000 years ago.

This ultimate type of social distancing is matched with a social problem-solving ability that is astute, for these hunter-gatherer societies engage in such ultimate social distancing only under special circumstances that make killing someone seem necessary. With relatively tractable deviants like Cephu, people take a chance on personal reform. But in their minds today's serial killers are likely to be tomorrow's serial killers, and there's only one sure way to deal with them.

In this context it's generally males who have to be socially eliminated, and these men are hunters whose contributions usually are beneficial to all. The obvious practical disadvantages inherent in losing hunters help to motivate band members to stop conflicts before they become homicidal and to refrain from using capital punishment if any other method will solve the problem at hand.

The possibility of reforming a deviant whose behavior is seriously perturbing group functions is helped by the fact that sympathetic feelings can play an important part in how such dilemmas are resolved. Feeling for the person makes it difficult to methodically execute any normal human being with whom social bonds exist. Thus, sympathy feeds into the universal prescriptions against murder that are found in all foraging societies and in human societies more generally—however "murder" may be locally defined. And feelings of sympathy also inform the reluctance of most moral majorities to use capital punishment at all, unless this is the only way out.

As they do their work in hunting societies, these small moral majorities are quite consistent in the ways they deal with social problems. Many deviants are treated as human beings who happen to be erring right now but are capable of reform, which is why with deviants like Cephu ostracism and shaming are used. It's the incorrigible bullies, serial killers, aggressively selfish or malicious shamans gone bad, and others who seriously threaten the lives or welfare or moral sensibilities of most or all group members, who are killed or expelled.

As we've seen, capital punishment affects a deviant's reproductive success profoundly, and the capital punishers have been kept in business in terms of natural selection because they are coming out ahead reproductively. Kill a major bully, and in an egalitarian group every member will, on average, share in the resources so gained. The same applies to a recidivist meat-cheater or an effective thief. Because the contest between good citizens and deviants is so often a zero-sum game, execution can pay off handsomely as far as the community is concerned—but so can reforming a lesser deviant who is contributing to the group economy.

In the preceding chapters, capital punishment of free-riding bullies or cheats has played an important part in our moral origins scenarios. The first scenario involved the appearance of a conscientious sense of right and wrong coupled with internalization of values, which explains moral origins without any Garden of Eden. It seems likely that the acquisition of a conscience as an effect of harsh social control involved not only physical elimination of individuals whose problems with lack of self-control were severe, but also lesser sanctions taking their toll on the fitness of lesser social deviants.

Once an effective conscience had evolved, the threat of severe group punishment had a major role in suppressing the potential predatory behaviors of free riders, and this brought quite a different effect. This development opened the way for modest yet socially significant human degrees of altruism to evolve genetically, bringing a greater degree of generosity to the operation of our consciences and to the overall tenor of our social lives.

Both of these developments were possible because ancestrally preadapted coalitions enabled our human ancestors to act as large groups in order to solve social problems. At first it was simply *political* majorities that did this, but once a conscience was in place, these majorities became *moral.*

LIFE UNDER A MORAL MAJORITY

In the minds of band members, social sanctioning comes from "the group," and it doesn't matter whether the entire group has reached a true consensus or whether a few people—often a deviant's kin—are staying neutrally to one side while the moral majority exerts its will.

Nor does it matter very much whether one or just a few individuals are stepping forward to act on behalf of the group; what counts is that they know they have the group—a moral majority—behind them. If they don't, an attempt to sanction can turn into simply a factional dispute, which degrades everybody's quality of social life and, as we've seen, likely splits the group.

Being able to arrive at a reasonably well-unified moral majority is critical to the continuing viability of any egalitarian band's social life, precisely because individual peacemakers are allowed too little authority to take on the tougher conflicts, and because some deviants are simply too awesome to be taken on without help or backing. These dynamics are understood by people in tiny bands that lack policemen, judges, and juries and often don't even recognize a single permanent headman.

A deviant who seriously irritates or threatens the majority of the band while at the same time violating a norm that has strong moral backing is definitely asking for trouble. The impending group reactions are easy enough to predict by an adult who has previously experienced such sanctioning—be it as a deviant or as a sanctioner. And simply participating in gossip is a constant reminder that group public opinion is, in general, coiled to strike. Such judgmental communities make the fabric of human social life moral, and they do so in predictable ways that have been in place for thousands of generations.

10 Pleistocene Ups, Downs, and Crashes

HUMAN NATURE AND HUMAN FLEXIBILITY

Culture is partly blind habit, but it also involves people solving problems. Because so much of social life is flexibly constructed in the light of conscious community needs and objectives, it should come as no surprise that as local environments change, people will be deliberately and insightfully modifying their practices to fulfill their needs as they see them.

It's time to consider exactly how morally flexible our culturally modern predecessors were likely to have been as they coped with radically changing prehistoric environments that could be inviting but also unpredictable and treacherous. We'll also be seeing how this flexibility ties in with the structurally ambivalent social nature we've evolved, for at times egoism can come into conflict with nepotism, while either egoism or nepotism can come into conflict with tendencies to be generous toward people outside the family. Our subject matter will be the sharing of food in times ranging from adequacy or excess to scarcity or outright famine.

HOW SHARING VARIES

In normal times, when the first wave of meat-sharing takes place with people like the Bushmen or the Netsilik or the forty-eight other LPA groups I've surveyed, we've already seen that through a variety of customs people behave as though the large carcasses were essentially the property of all—at least until the meat is distributed. This group appropriation of large carcasses is sustained largely by pressure from public opinion, but if necessary, brute physical force can be employed to prevent monopolization that favors individuals or their families.

In combination, social pressure and active sanctioning do quite a good job of keeping the self-aggrandizing tendencies of egoistic or overly nepotistic hunters in check—especially when they are the ones who have proudly brought home the bacon and are feeling their oats. The immediate winners of this zero-sum game are all the other members of the band, and the underlying idea is that the band is one big cooperative team as far as large game is concerned.

This pattern applies to normal times when killing large game is challenging and quite unpredictable in its acquisition, yet plentiful enough to be extremely useful to group subsistence. The result is the efficient, culturally routinized type of meat-sharing system that we've discussed at such great length. Not all times are normal, however, and in the Pleistocene Epoch this must have become gruesomely apparent again and again, as environments that sometimes were benevolent often changed drastically for the worse and presented our ancestors with threatening scarcities worthy of the term "dire."

VARIETIES OF SHARING IN GOOD TIMES

To begin with, however, we must ask what happens in those rare times when seasonal resources are so ample, and so highly concentrated, that a cooperating population can congregate at a great size such that the usual system of meat-sharing simply becomes unwieldy. In the far north of central Canada, the Netsilik enjoy this possibility every year, and we'll now take a closer look at their remarkable system for averaging out shares of seal carcasses, which involves a rather fancy, large-scale version of Alexander's indirect reciprocity. Just in the winter, a sizable number of small bands gather together on the sea ice to live well on seals for several months, and these groupings can exceed sixty to eighty persons. Hunting seals through their multiple breathing holes is accompanied by a high degree of unpredictability, and the Netsilik have created a brilliant system of meat distribution that reduces family meat-intake variance just as we've seen in small, normal-sized bands.[1]

If we imagine a butcher's chart with a picture of a seal divided into seven major parts, then we're looking at the main basis for this system. When any participating hunter lands a seal, he'll have a different permanent partner for each body part; for instance, he'll always give the flippers to his flipper partner—and that hunter will reciprocate in kind when he's lucky enough to land a seal.[2] With a system like this, everyone's eating some seal meat all of the time. And seven hunters would be right in the ballpark for the number of meat-procurers and meat-sharers in a normal band of thirty.

It obviously wouldn't make much sense for sixty people to try to share a single seal. But what would be the alternative if all these Netsilik families simply hunted on their own? With a little bad luck, even an adept hunter and his near kin might kill two seals one month but then none the following month. So this system of "meat insurance" sees to it, in terms of probabilities, that all seven of these sharing partners will eat adequate, if moderate, amounts of meat and blubber quite regularly, thereby achieving variance reduction. All of these partners will be nonkinsmen; within their extended families, Inuit kinsmen share automatically, so they don't need such a system to regulate their meat intake.[3]

A very different kind of cultural flexibility obtains in times of truly glorious surfeit. When people are living in normal bands and major windfalls occur, the fair sharing between unrelated families can simply disappear, as when other Inuit groups intercept large caribou herds at narrow river crossings and slaughter scores of animals.[4] With so much meat, sharing doesn't make any sense until the meat supply becomes sporadic again. At that point, the normal variance-reduction system will cut in again to effectively average out family meat intake.

THE CONTENTIOUS SIDE OF SHARING

There's an opposite side to this coin, which arrives when encounters with game become unusually rare. However, before we examine what transpires when foragers are faced with dire scarcity, the matter of motives needs some further clarification. Basic motives that affect patterns of sharing are likely to be complicated because, as we've seen, they're likely to be mixed. There's motivationally generous sharing, done simply because the need of a socially bonded other is sensed and sympathetically appreciated; this other person may be someone very familiar or merely an acquaintance or at times even simply a member of the same human race. There's also sharing because of a deeply internalized sense of duty to share. And then there's sharing to avoid shame and gain reputation, and there's sharing to be eligible for future benefits. The following discussion will help in clarifying not only this complex motivational division of labor, but also what the limits of hunter-gatherer generosity may be.

Let's reconsider what happens after a normal LPA meat distribution has taken place in a thriving band of twenty to thirty. Once it's shared out, we know the meat amounts to private property. This means that giving to others outside the family becomes optional, whereas sharing within the family is basically automatic even though there may be some jealousy or squabbles. We also know that stinginess is poorly thought of and that generosity is promoted in general. We also know that, optionally, private shares of meat can be shared between unrelated families—even though some of the sharing is likely to be ambivalent because underlying egoistic and nepotistic tendencies are working against competing underlying tendencies that favor extrafamilial generosity.

Rarely, such ambivalences can spawn serious social conflict over meat, for instance, if requests thought to be legitimate are outright denied and expected informal reciprocation is not forthcoming spontaneously. Such problems are recorded for both the Inuit[5] and in the Kalahari,[6] and Niqi, the woman with the problematically stingy personality in Briggs's Utku band, found herself being partially ostracized because she held back.[7] Surely, such problems can be generalized to LPA foragers in general. But this includes the fact that in this context of informal meat-sharing between families really serious quarrels—ones that threaten to split the band—are very rare.

We've already met with the well-known school of thought that likens hunter-gatherer meat-sharing to "tolerated theft" as this has been modeled for chimpanzees, and this somewhat "cynical" model does seem to go nicely with certain facts about human sharing. First, in some but by no means the majority of LPA foraging societies, we've seen that the initial, "communalized" wave of sharing can be accompanied by routine and constant demands for more meat.[8] These characterizations amount to indirect or direct accusations of unfairness in terms of customary rules, and perhaps the underlying hostility could be seen as threatening enough to motivate sharing. However, even in groups with contentious *styles* of sharing, large carcasses are always shared so that everyone gets some meat regularly, and even in such bickering groups there is pleasure in eating together and—to repeat—*serious* quarrels are very rare.

Although these contentious styles of sharing vary greatly in their intensity, I believe they all can be viewed similarly to how we have depicted the universal pronouncements that praise extrafamilial generosity. They amount to culturally routinized ways of "tweaking" other people's relatively modest tendencies to behave generously outside the family, and the tweakers are merely tweaking; they're not spoiling for a fight that will, at cost to everyone, split the band.

Such deliberate amplification of extrafamilial generosity is evident not only when nagging is used to spur generosity or when groups issue the explicit calls for generosity that we've discussed at length,[9] but also less directly when they punish cheating hunters or sneaky thieves who seriously break the rules of sharing. All of these manipulative measures suggest that sharing this precious commodity is accomplished in the face of considerable ambivalence.

One important take-home message is that, even though at the level of genetic dispositions altruism is obviously outranked by egoism and nepotism, the feelings of care and generosity that promote altruism still can reduce the potential for conflict because basically they take some of the edge off of individual competition for food. In effect, I believe that human generosity oils the wheels of cooperation and that people understand this intuitively; that's why generous tendencies are constantly being tweaked. The result can be good for everyone. In fat or adequate times, this makes it possible for people to cooperate quite efficiently—and take pleasure in doing so in spite of the occasional rough edges we've discussed. Thus, today's various negative and positive pressures to be generous toward nonkin would have been a very important component of the sharing systems that were in effect 45,000 years ago among culturally modern humans, and probably earlier as well.

Today, both the !Kung and the Netsilik share routinely within their families, and in both cultures the people sometimes know hunger. However, the !Kung seem to be one of the world's more contentious cultures—not only when it comes to initial sharing of communalized meat at the band level, but also afterward when private meat is requested by members of other families. In this connection I should mention that in general the !Kung seem to be unusually loquacious and also perhaps more prone to heated interpersonal conflict—even though they do praise generosity and try hard to reduce conflict just like all other LPA foragers.

I should also say that in spite of a contentious verbal style, the !Kung do seem to share quite effectively, just as Pat Draper told us they do. What may be distinctive of the !Kung and certain Australian

Aborigines,[10] and a handful of others whose styles of sharing are unusually contentious, is that they don't try to hide their ambivalences and their worries about fair apportionment of meat, as the Inuit and many other hunter-gatherers seem to. In my view, the more contentious cultures are simply making overt anxieties about fair sharing that are widespread, and they are widespread because egoism and nepotism can so readily trump human altruism.

HOW GENEROUS FEELINGS SUPPORT COOPERATION

In thinking about human nature, here's my thesis. Critically important are the underlying generous feelings that help a system of indirectly reciprocated meat-sharing to be invented and maintained. Yet it's also true that basically these altruistic tendencies are so moderate that hunter-gatherer sharing institutions need continuous and strong positive cultural support if cooperative benefits are to be reaped without undue conflict. In a sense, then, these innately generous tendencies are not quite up to the job. To finish the job at the cultural level, the serious and continuous threat of group disapproval and active sanctioning does its part in making systems of indirect reciprocity among nonkin work without too much conflict. So does a simple awareness that living in sizable bands pays off and that serious conflict will cause the band to split. And so do all the positive preachings in favor of generosity, and so, sometimes, does hostilely manipulative nagging.

If all of this tweaking of the better side of human nature is needed, we might ask to what degree such sharing is emotionally "genuine" in that it is based importantly on actual feelings of generosity. This surely varies by the personality of the sharer, by the situational context of sharing, by the degree of bonding between sharers and recipients, and by whether kin are involved. But in normal times people do share and they do so quite efficiently, and this is usually the pattern—in the Holocene. But what about the Late Pleistocene Epoch, when climates were so capricious—and when frequently they could be cruel and even highly dangerous?

DIRE SCARCITY BRINGS THE END OF SHARING

When we think about this division of labor among egoism, nepotism, and wider generosity, what takes place when food becomes really scarce is of great interest. Keep in mind that when large game has become very hard to acquire, and is seriously emaciated once found, usually the climatic conditions that have this effect will also be reducing the chances of humans subsisting well on plant foods and small game.

In the Arctic, of course, there simply is no such vegetarian fallback strategy. There, freezing is a major means of storage, but there are obstacles to its use as well; for instance, other carnivores, including well-motivated wolves and powerful bears, can raid these caches even if there are stones available to place over the meat. There's also the energetic problem of digging through the permafrost if the cache is to be deep. People like the Netsilik do sometimes freeze some surplus meat, but as we'll see, this won't be enough to get them through an unusually hard winter.

Archaeologist Lawrence Keeley surveyed forty North American foraging societies for which information on food scarcity was available,[11] and a quarter of them were what I've been calling LPA bands, mostly Inuit. Thirteen societies commonly experiencing famines with people dying of starvation included Inuit-speakers living in inland areas of Canada, such as the Netsilik, and a number of inland-dwelling boreal-forest Athapascans in the Pacific Northwest, non-LPA foragers who were engaged in the fur trade. Subject to *occasional* famines were another twelve non-LPA fur-trade groups that foraged in inland subarctic forests. The fifteen societies that were famine free mostly lived in California and were sedentary. Keeley did not include mounted hunters of the Great Plains in his sample.

One bottom-line finding was that some areas make for a steady, dependable living, whereas others present resources that fluctuate so much over time that years with famines are predictable. It's of interest that the famine-free groups were in California, where resources were ample enough and concentrated enough so that most of the foragers could live in groups larger than the usual nomadic bands, basically stay in permanent settlements, and store their food as families. This would have been a deviation from the basic nomadic pattern before 15,000 years ago, although such outliers probably existed then.

In North America, hunger also struck people living in the dry and erratic Great Basin, an area that can be compared to the Kalahari, and to parts of the desertic interior of Australia, where similar food shortages also are known.[12] But it's no accident that our main accounts of hunter-gatherer famine come from the central and eastern Arctic and from subarctic boreal forests because plant foods (or small game or insects) offer so little backup when the meat supply fails.

Once famine is on its way, sharing appears to break down in two stages. In coming to this conclusion, I'm keeping in mind the rather slim hunter-gatherer data,[13] but also the more abundant information we have about agricultural tribesmen who live in similarly small groups and in the context of drought sometimes face starvation[14] in spite of their using long-term food storage.[15] With respect to bands, as famine approaches, the various families may continue to share any plant foods and small game within the family—just as they do in times of plenty. However, large-game consumption will have been diminishing for three reasons: game becomes scarce, hungry hunters have less energy to hunt, and when killed, emaciated large mammals are providing much less meat. Furthermore, as variance increases, sharing at the band level begins to make less sense for statistical reasons.

In this context of nutritional downturn, predictably nepotistic tendencies will tend to decisively trump altruistic tendencies to share with nonrelatives. And when privation turns into famine, a quick glimpse tells us that what can happen next isn't at all pretty. Within family circles, underlying egoistic tendencies can start to trump nepotistic tendencies, and even within close nuclear families, sharing can be drastically reduced. In extreme situations even intrafamilial cannibalism can take place, with the main examples coming from the Arctic.[16] However, our very powerful egoism, our strong nepotism, and our relatively modest altruism predict very much the same behavior elsewhere when hunter-gatherers face an imminent starving-to-death situation.

What made for privation in Pleistocene Africa wouldn't have been the absence of plant foods as a fallback, but periodic abrupt and radical onsets of droughts that would have affected plants and animals alike. Hard times surely appeared far more often than recently, and in recurrent cycles, and for humans highly flexible responses to the opportunity to share would have provided major adaptive advantages in dealing with these regional climate swings, which within a half a century or less surely could have moved environments from bountiful to dangerously inadequate or even worse.

Long ago, an agronomist named Justis von Leibig promulgated his environmental "Law of the Minimum,"[17] which suggests that in normal times the effects of natural selection on gene pools may not be nearly as forceful as when certain resources become very scarce and therefore limiting. Thus, in the far north mammals and fish are the limiting factor for humans because snow and ice provide water and plants have little relevancy. On the Kalahari, availability of water could be the main limiting variable.

Combine this insight, now, with the fact that prehistorically humans seem to have experienced bottlenecks during which the overall number of people in existence became quite small.[18] When both plant foods and animals were subject to short-term failure, local-population extinctions were very likely,[19] and on at least one prehistoric occasion total species extinction was a real possibility.[20] In this context advantages of the very wide kind of flexibility in sharing we have been speaking of could have made a major difference to bands or families or individuals, as they made often very ambivalent decisions about how and when to share.

It's heartbreaking to think of how often during the Late Pleistocene culturally modern beings with moral sensibilities like ours were faced with extreme and cruel situations of triage. Over and over again they were obliged to set aside deeply internalized moral values that favored helping others—be this through altruism or nepotism—in the interest of family or even personal survival. When the chips were down, band-level sharing became all but absent, which means that hungry families and individuals were very much the units of selection because the band as a whole had ceased to cooperate. And even family-level cooperation was up for grabs: in some contexts it could have paid off reproductively for parents to share scarce foods with offspring or with each other; in more pressing circumstances it would have made sense in terms of survival not to share, which made individuals the units of selection. The ultimate dilemma, faced historically by the Netsilik in central Canada, was whether to turn cannibal. In one case,[21] parents who ate their own children rationalized their actions as follows: had the parents killed themselves for the children to eat, the children still couldn't have made a living by hunting when the wild game returned. However, if the parents survived by consuming their offspring, they could at least have more children and thereby keep things going.

This brings us back to the human conscience and the adaptiveness of its flexibility. A too-strong conscience would make such stressful adaptive moves unthinkable. However, a flexible conscience allowed people to adjust their adherence to moral rules to the situations they faced, and when altruistic empathy was trumped by egoism or nepotism, apparently they were able to do what was necessary.

Let me suggest exactly how as hunger escalated, and variance in meat intake increased, having a flexible, as opposed to a rigid, conscience would have helped to make these facultative responses possible. First, contingent sharing with nonkin in the same band was likely to be too immediately costly and it ceased to make sense because the potential sharer might be dead before possibly very eventual future contingencies came into play. Second, further triage would have phased in when sharing began to break down within the family so that even innately strong, culturally supported nepotistic impulses could be trumped by desperately hungry decisions in favor of egoism. The ultimate victory of egoism took place when cannibalism on nearby dead bodies came into play and even more so when the active killing and eating of others by people too weak to forage probably took place.[22]

If we consider this entire range of solutions to the general problem of sharing food, the interplay among egoism, nepotism, and altruism is obvious enough. Also obvious are cultural emphases that strongly promote generosity to both kin and nonkin but that in practice fall short of being absolute. Apparently, cultural values can even condone sheer egoism in the face of famine, or so it seems when practical Inuit individuals known to have resorted to intrafamilial cannibalism appear not to have been socially punished afterward.[23]

AMBIVALENT SHARERS

Let us return to some contemporary foragers who on an everyday basis can't even live in bands. They live so close to *routine* scarcity that the usual highly interdependent multifamily band life all but disappears for long stretches of time. This takes place owing to necessity out on certain marginally productive Australian semideserts, where game is scarce and insects and lizards are nutritionally important. Likewise, in the dry Great Basin of northwestern North America, with just a few deer available, Shoshonean family groups have to depend on highly unpredictable pine-nut harvests, and a limited rabbit supply, when it comes to getting some protein and fat.[24] In such situations, sharing within families can stay in effect, and the dispersed families will still meet when they can to socialize (and share meat) on a larger scale. If there's a large pine-nut harvest, they may even be able to form sedentary bands for up to a year.[25] But nutrition comes first, and over the years, their much beloved band-level social life could all but disappear for much of the time.

Because of our love of company, the human social preference is to live in bands, not just in families, but we've just seen that foragers can adjust. This means that in the Pleistocene periodically marginal but viable subsistence adjustments would have led to dispersed families barely getting through temporary hard times, when staying bunched up in a band would have made staying alive much more uncertain because bands used up the limited food around them so quickly compared to families.

Over a decade ago, archaeologist Rick Potts took a look at the startling incoming data on Pleistocene climatic instability and decided that adaptive *flexibility* must have been an important key to human survival, a key so important that our brain's remarkable advances in size could be attributed largely to frequent and challenging environmental changes and the need to cope with them intelligently.[26] Here I'm combining Potts's hypothesis and Leibig's Law with hypotheses about our inherently ambivalent human nature, and with our morally flexible ways of dealing with scarcity, to suggest that important, punctuated phases of human genetic evolution must have taken place when people were very hungry—and some of them simply weren't going to make it.

In prehistoric hard times the overall mix of biocultural components I'm describing here provided a remarkably flexible way for humans to at least stay alive as small kin groups—or even just as individuals. Thus, these units became *highly* relevant as units of selection, while multifamily groups became temporarily less relevant unless there was active conflict over defensible resources. At the same time, when resources were adequate, this same behavioral potential provided a way for humans to really flourish in bands as they gradually built up their populations before the next crisis struck.

Of course, this flexibility's advantages for adaptation have also been useful more recently in the far more stable—and usually more comfortable and predictable—Holocene Epoch. But under Pleistocene conditions, at frequent junctures such adaptive flexibility surely was far more critical, much more often, as entire regions became "marginalized" and people had to cope with dire hunger and had to curb their culturally reinforced generosity.

Sometimes I wonder if these once quite frequent environmental shortfalls, and hunter-gatherers' flexible responses to them past and present, have been adequately exposed to scientific analysis. When anthropologists write their ethnographies, the daunting task of describing a culture holistically usually results in a "normalized" depiction of social life and sharing that doesn't fully take into account these more unusual emergency contingencies,[27] even if they are remembered by older people. In addition, archaeological evidence for *temporary* dire scarcity, and specifically its social effects, is by its nature rather slim. For these reasons, we must not idealize today's hunter-gatherers, with their routine sharing of large-game meat as a prized and nutritiously important commodity, by looking only to the good-times accounts that predominate in ethnographic reports.

Even in good times, perfectly motivated, all-but-automatic generosity is not the name of the sharing game, especially outside the family. I've described the underlying ambivalences that can surface actively, and I've emphasized that our helpful altruism becomes socially potent only when this potential is culturally boosted through constant calls for generosity, praise for the generous, and criticism and social punishment for the exceptionally stingy. Of course, one way an outsider might read all the prosocial messages is that the people in question are just naturally very generous and cooperative, and golden rule sayings might seem to reflect this. Another, however, is that hunter-gatherer systems of sharing are so fragile that they need very substantial and continual reinforcement.

This "battle" is not entirely uphill. From an early age these same people have internalized cultural values that promote altruistic generosity and therefore are naturally responsive to these prosocial messages. This must not be overshadowed by the various rough edges I have described, as with all the !Kung complaints about meat-sharing. The latter are simply expectable manifestations of egoism and nepotism at work in cultural traditions that allow their strong expression. The much less contentious-seeming Netsilik are probably closer to the worldwide LPA forager norm,[28] but even though they may not use repeated demands to get their share, I can assure you that Netsilik women or men who are saddled with the difficult task of sharing out the meat fairly are watched closely, and resentments over failure to share can be bitter. This would be true anywhere that highly valued and not overly plentiful large game has to be shared, and with respect to the !Kung I have suggested that they provide a good idea of how a system of sharing can be highly efficient in spite of peoples' worried anticipation of stinginess.

NISA

In dealing with this interplay between human genetic nature and these cultural patterns that amplify generosity, I wish to get up close and personal again. As we've seen, the !Kung are probably quite near to the complaining extreme among today's LPA foragers in general.[29] In this context, the words of Nisa, the !Kung woman whose autobiography will provide us with further detailed insight into how the !Kung feel about things like sharing, will enable us to vicariously enter one Bushman mind and consider the underlying psychological stress at first hand.

Nisa comes across as a classically ambivalent sharer, and it started young. Of course, her earliest recollection of a problem in this area came when she was being weaned after her mother became pregnant. It's usual for hunter-gatherers to continue nursing until a child is several years old, and because lactation tends to suppress ovulation, this provides a long birth interval that is congenial to a nomadic way of life—where carrying two very small children would be a serious burden on all concerned. I shall hazard a guess that in our own culture very few people can remember being weaned, but when maternal warmth and nourishment were abruptly denied to Nisa as a child already several years old, she appears to have suffered a real and well-remembered trauma. This caused her to engage in parentally disapproved behavior both before and after the birth of her younger sibling.

Marjorie Shostak's remarkable account is replete with both childhood and adult remembrances of being angry with others for being stingy—with Nisa tearfully or angrily wanting to retaliate in kind. In fact, save for one episode that involved her being unduly generous, if we were to generalize just Nisa's personally remembered behavior patterns to all Bushmen, we would have to wonder how they managed to share at all.

This is an occupational hazard when working with a single autobiography, and there are also our own ethnocentric reactions to deal with. I probably could not rely on such an account were it not for the accompanying insights of the ethnographer. Indeed, Shostak tells us that other Bushman children have similar problems: "The !Kung economy is based on sharing, and children are encouraged to share things from their infancy. Among the first words a child learns are *na* ('give it to me') and *ihn* ('take this'). But sharing is hard for children to learn, especially when they are expected to share with someone they resent or dislike. And giving or withholding food or possessions may be a powerful way to express anger, jealousy, and resentment, as well as love."[30]

Although Nisa's actual food jealousies may well fall within the normal range, it does appear that her memory seems to work quite well when she's recalling past deprivations and conflicts. Shostak continues, "It is also hard to learn not simply to take what you want, when you want it. !Kung children rarely go hungry; even in the occasional times when food is scarce, they get preferential treatment. Food is sometimes withheld as a form of punishment for wasting or destroying it, but such punishment is always short-lived. Nevertheless, many adults recall 'stealing' food as children. These episodes reflect the general !Kung anxiety about their food supply, as well as the pleasure they take in food—both emotions already present in childhood."[31]

Nisa's autobiography is replete with stories involving food:[32]

> When mother was pregnant with Kumsa, I was always crying, wasn't I? I would cry for a while, then be quiet and sit around, eating regular food: sweet nin berries and starchy chon and klaru bulbs, foods of the rainy season. One day, after I had eaten and was full, I said, "Mommy, won't you let me have just a little milk? Please, let me nurse." She cried, "Mother! My breasts are things of shit! Shit! Yes, the milk is like vomit and smells terrible. You can't

drink it. If you do, you'll go, 'Whaagh . . . Whaagh . . . ' and throw up." I said, "No, I won't throw up, I'll just nurse." But she refused and said, "Tomorrow, Daddy will trap a springhare, just for you to eat." When I heard that, my heart was happy again.

The next day, my father killed a springhare. When I saw him coming home with it, I shouted, "Ho, ho, Daddy! Ho, ho, Daddy's come! Daddy killed a springhare; Daddy's bringing home meat! Now I will eat and won't give any to *her*." My father cooked the meat and when it was done, I ate and ate and ate. I told her, "You stinged your milk, so I'll stinge this meat. You think your breasts are such wonderful things? They're not, they're terrible things." She said, "Nisa, please listen to me—my milk is not good for you anymore." I said, "Grandmother! I don't want it anymore! I'll eat meat instead. I'll never have anything to do with your breasts again. I'll just eat the meat Daddy and Dau kill for me."

On one occasion, Nisa refuses to share with her mother some duiker meat that was acquired by her older brother, Dau:

After he skinned it, he gave me the feet. I put them in the coals to roast. Then he gave me some meat from the calf and I put that in the coals, too. When it was ready, I ate and ate and ate. Mother told me to give her some, but I refused, "Didn't you stinge your breasts? Didn't I say I wanted to nurse? I'm the only one who's going to eat this meat. I won't give any of it to you!" She said, "The milk you want belongs to your brother. What's making you still want to nurse?" I said, "My big brother killed this duiker. You won't have any of it. Not *you*. He'll cut the rest into strips and hang it to dry for me to eat. You refused to let me nurse so your son could. Now you say I should give you meat?"

The following passages tie a jealous Nisa to stealing her mother's milk through deceit, and in the end it is her father who lays down the law with threats of corporal punishment:

Another day, my mother was lying down asleep with Kumsa, and I quietly sneaked up on them. I took Kumsa away from her, put him down on the other side of the hut, and came back and lay down beside her. While she slept, I took her nipple, put it in my mouth and began to nurse. I nursed and nursed and nursed. Maybe she thought it was my little brother. But he was still lying where I left him, while I stole her milk. I had already begun to feel wonderfully full when she woke up. She saw me and cried, "Where . . . tell me . . . what did you do with Kumsa? Where is he?" At that moment, he started to cry. I said, "He's over there."

She grabbed me and pushed me, hard, away from her. I lay there and cried. She went to Kumsa, picked him up, and laid him down beside her. She insulted me, cursing my genitals, "Have you gone crazy? Nisa-Big-Genitals, what's the matter with you? What craziness grabbed you that you took Kumsa, put him somewhere else, then lay down and nursed? Nisa-Big-Genitals! You must be crazy! I thought it was Kumsa nursing!" I lay there, crying. Then I said, "I've already nursed. I'm full. Let your baby nurse now. Go, feed him. I'm going to play." I got up and went and played. . . .

Later, when my father came back from the bush, she said, "Do you see what kind of mind your daughter has? Go, hit her! Hit her after you hear what she's done. Your daughter almost killed Kumsa! This tiny little baby, this tiny little thing, she took from beside me and dropped somewhere else. I was lying down, holding him, and fell asleep. That's when she took him from me and left him by himself. She came back, lay down, and started to nurse. Now, hit your daughter!"

I lied, "What? She's lying! Me . . . Daddy, I didn't nurse. I didn't take Kumsa and leave him by himself. Truly, I didn't. She's tricking you. She's lying. I didn't nurse. I don't even want her milk anymore." My father said, "If I ever hear of this again, I'll beat you! Don't ever do something like that again!" I said, "Yes, he's my little brother, isn't he? My brother, my little baby brother,

> and I *love* him. I won't do that again. He can nurse all by himself. Daddy, even if you're not here, I won't steal mommy's breasts. They belong to my brother." He said, "Yes, daughter. But if you ever try to nurse your mother's breasts again, I'll hit you so that it *really* hurts." I said, "Eh, from now on, I'm going to go wherever you go. When you go to the bush, I'll go with you. The two of us will kill springhare together and you'll trap guinea fowl and you'll give them all to me."

Subsequently, Nisa turns into a childish thief whose anxious orientation to feeding continues to get her in trouble in terms of a local sharing ethic that, as my survey showed, emphasizes being generous both within the family and without. And within the family, her own childishly "deviant" patterns are treated sternly by her mother in particular:

> This was also when I used to steal food, although it only happened once in a while. Some days I wouldn't steal anything and would just stay around playing, without doing any mischief. But other times, when they left me in the village, I'd steal and ruin their things. That's what they said when they yelled at me and hit me. They said I had no sense.
>
> It happened over all types of food: sweet nin berries or klaru bulbs, other times it was mongongo nuts. I'd think, "Uhn, uhn, they won't give me any of that. But if I steal it, they'll hit me." Sometimes, before my mother went gathering, she'd leave food inside a leather pouch and hang it high on one of the branches inside the hut. If it was klaru, she'd peel off the skins before putting them inside.
>
> But as soon as she left, I'd steal whatever was left in the bag. I'd find the biggest bulbs and take them. I'd hang the bag back on the branch and go sit somewhere to eat. When my mother came back, she'd say, "Oh! Nisa was in here and stole all the bulbs!" She'd hit me and yell, "Don't steal! What's the matter

with you that inside you there is so much stealing? Stop taking things! Why are you so full of something like that?"

One day, right after they left, I climbed the tree where she had hung the pouch, took out some bulbs, put the pouch back, and mashed them with water in a mortar. I put the paste in a pot and cooked it. When it was ready, I ate and finished everything I had stolen.

Another time, I took some klaru and kept the bulbs beside me, eating them very slowly. That's when mother came back and caught me. She grabbed me and hit me, "Nisa, stop stealing! Are you the only one who wants to eat klaru? Now, let me take what's left and cook them for all of us to eat. Did you really think you were the only one who was going to eat them all?" I didn't answer and started to cry. She roasted the rest of the klaru and the whole family ate. I sat there, crying. She said, "Oh, this one has no sense, finishing all those klaru like that. Those are the ones I had peeled and had left in the pouch. Has she no sense at all?" I cried, "Mommy, don't talk like that." She wanted to hit me, but my father wouldn't let her.

Another time, I was out gathering with my mother, my father, and my older brother. After a while, I said, "Mommy, give me some klaru." She said, "I still have to peel these. As soon as I do, we'll go back to the village and eat them." I had also been digging klaru to take back to the village, but I ate all I could dig. . . . Later, I sat down in the shade of a tree while they gathered nearby. As soon as they had moved far enough away, I climbed the tree where they had left a pouch hanging, full of klaru, and stole the bulbs.

This childishly antisocial eating spree brings on physical punishment:

"Nisa, you ate the klaru! What do you have to say for yourself?" I said, "Uhn, uhn, I didn't take them." My mother said, "So, you're

> afraid of your skin hurting, afraid of being hit?" I said, "Uhn, uhn, I didn't eat those klaru." She said, "You ate them. You certainly did. Now, don't do that again! What's making you keep on stealing?"
>
> My older brother said, "Mother, don't punish her today. You've already hit her too many times. Just leave her alone."
>
> "We can see. She says she didn't steal the klaru. Well then, what did eat them? Who else was here?"
>
> I started to cry. Mother broke off a branch and hit me, "Don't steal! Can't you understand! I tell you, but you don't listen. Don't your ears hear when I talk to you?" I said, "Uhn, uhn. Mommy's been making me feel bad for too long now. I'm going to go stay with Grandma. . . . I'm going to go stay with Grandma. I'll go where she goes and sleep beside her wherever she sleeps. And when she goes out digging klaru, I'll eat what she brings back."

Nisa proves to be too much for her grandmother, who is too old to do much gathering. And her obviously intense problems with food envy or delay of gratification seem to have persisted into adulthood a bit more, *perhaps,* than is typical of her !Kung Bushman peers. But to the large extent that she seems to be typical, her behavior reflects quite nicely the underlying food anxieties of the Bushmen and the expressive manner in which mature Bushmen, who in fact do encounter periodic food scarcity, vociferously resent any anticipated lack of sharing on the parts of others.

Curiously, in the interviews Nisa never chooses to emphasize her own generosity. As an adult, Nisa's style of narration suggests that these childish patterns continued, but it's difficult to determine the degree to which her constant complaints, voiced in these very private recorded sessions with Shostak, fairly represent what was really going on in Nisa's social life. For instance, we have the adult Nisa actually receiving complaints from her husband that she herself is much too generous. This generous side appears in passing, almost like the tip of an iceberg, while Marjorie and Nisa are discussing Bo, Nisa's husband. Marjorie starts things off:

> I asked, "And your hearts, do they go out toward each other?" She said, "Yes, our hearts love each other and go out toward each other." What about fights? "We rarely fight. When we do, it's usually about food, when I serve too many people. That's when he asks, 'What are you doing, serving everyone? When do others ever serve us? When we have food, we should be the only ones to eat.' But I say, 'You just like to yell about things.' Then he says, 'It's because you are bad, a bad one that sees a person and gives him food, then sees another and gives him food. Don't you know that when you have food, it is for you and your child, that she can eat and be full? You'll wind up just like a woman with nothing this way.'" Was this an important fight? "No, it's very small. We fight a little, then leave it and love each other again."[33]

Can a somewhat idiosyncratic indigenous autobiography be useful to evolutionary analysis? A flattering presentation of self would be one predictable distortion, but in fact Nisa doesn't seem to be doing much of this, perhaps because Marjorie is her trusted confidante. It would be useful to have dozens of these personal accounts so that central tendencies in Bushman personality patterns would be easy to identify, and it would be very interesting to have for comparison another dozen from an LPA society where people are seldom hungry or are less prone to nag others about meat-sharing. But we're lucky that Marjorie Shostak went to the trouble of eliciting and transcribing Nisa's personal account, for the hints it provides are important.

ADAPTATION

In fact, Nisa's ambivalences seem to reflect generalized Bushman ambivalences that are quite predictable. First, consider the fact that these people often experience extensive routine privation in the lengthy dry season and that sometimes because of localized minidroughts they must travel long distances to keep from starving.[34] Then consider the fundamental underlying conflicts discussed previously, which stem from having a genetic nature that is only moderately altruistic and a conscience that is morally flexible. The social solutions that these and other LPA foragers like the Netsilik have come up with tell us a great deal about how culturally modern humans made it through the Late Pleistocene, with its radical ups and downs, and into a Holocene land of plenty, where even many "marginalized" environments were at least fairly steady.

We may assume that even after childhood learning experiences are complete, the internalization of nepotistic values that favor generosity within the family will not begin to wholly neutralize strong egoistic drives when they combine with extreme hunger. The biological dispositions that help people learn to be generous to *non*kin are weaker still, which is why when famine begins to approach, and in spite of the band-wide sharing system's being well reinforced by moral beliefs when times are normal, this system as a whole will begin to fall apart.

Another way of saying this is flexibility. These foragers have built social edifices that are influenced by a hierarchy of competing motivations and also by a range of different environmental conditions. And institutions like meat-sharing reflect practical concerns about when profit is to be had in engaging in a system of band level indirect reciprocity and when such participation should be set aside. Given the relatively limited motivations that natural selection has given us to work with for being generous outside of the family, it sometimes strikes me as remarkable that we share as well as we do. The cultural amplification of these modest but very important altruistic tendencies provides much of the answer to the question as to how such institutions can be maintained, and we must give credit to the socially sophisticated minds that helped these language-assisted systems of cultural reinforcement to develop in the first place.

On a day-to-day basis these same minds understand established indigenous sharing systems, and when they are profitable, people strive to make them work well on a continuing basis because in normal times this significantly raises their standard of living. Of course, the same minds shape quite different adaptive responses in times of unusual abundance, when band-level systems of sharing simply become superfluous. Finally, when dire scarcity strikes, people's decisions can become downright cruel, as opposed to generous, even though they may still experience pangs of sympathy and acutely feel the moral compromises they must make.

SMART ENOUGH TO BE FLEXIBLE

This chapter has explored the day-to-day expression of cooperative generosity between families and the potentially formidable obstacles to it that exist in the form of egoism and nepotism. In the Late Pleistocene, it was being moral that enabled relatively weak dispositions to altruism to make humans as cooperative as they often were able to be. Conscience-based internalization was important—the same internalization of values that made the adult Nisa a person who apparently was capable of overgenerosity even though in a situation of privacy and trust she resentfully complained all the time to Marjorie Shostak. Also important was the deliberate social reinforcement of extrafamilial generosity that I've mentioned so often. This type of giving was praised by moralistic band members who looked down on stinginess—and were prepared to punish bullies or cheaters, if these really serious free riders sought outsized shares of scarce desirables when the usual system of indirect reciprocity was operative.

Cognitive capacity was important, for our culturally modern forbears were likely to have known exactly what they were doing socially when they reinforced people's generosity and curtailed major free riders. In coping with the special exigencies they faced in partially pursuing a social carnivore's subsistence strategy, this culturally based capacity for "social engineering" enabled them to develop efficient systems of meat-sharing that worked extremely well in times of adequacy, and also, I believe, worked reasonably well when they faced times of marginal but not dire dietary inadequacy.

This same capacity to make strategic decisions also enabled people like the hungry Netsilik to reject their own customary sharing practices when food became so scarce that trusting in a long-term system of indirect reciprocity became life threatening. At that point, the group social control that kept such systems going would simply fade away, while cooperation within the family became still more important—unless a really dire famine presented itself. In that case, social acts that otherwise would have been punishably monstrous were apparently "understood" by fellow moralists.

Our "parliament" of competing instincts[35] was being mediated by an evolutionary conscience, which did permit a total cessation of sharing when *this* made sense. Like individual consciences, group moral beliefs also were flexible: for instance, they did not call for the punishment of unfortunate cannibals who in dire straits were obliged to either eat their fellows—or die. Thus, for Late Pleistocene humans, doing unto unrelated others worked quite well when times were decent for subsistence, as they have been in so many places during most of the Holocene. But over and over again in the capricious Pleistocene, a profound degree of flexibility was needed as culturally modern humans like ourselves scrambled to survive as they faced critical shortages of meat, plant foods, or water.

Today, the scope of human generosity is still highly adjustable—precisely because if we had not been able to evolve in that direction, we might well not have survived earlier on. To Americans who go to supermarkets for their food, Nisa's gastronomic anxieties may seem obsessive. But they are a manifestation of a culture that has become responsive to periodic privation.

In the Pleistocene, people everywhere faced far worse and surely quite frequent crises as their capricious environments changed for the better and then for the worse and often became outright untenable. Today's LPA foragers rarely have to cope with repeatedly abrupt and lasting ups and downs like those faced by foragers who had to cope with the Pleistocene Epoch, but the mix of human capabilities that enabled those earlier human beings to stay in business was evolved by them, not by us. The adjustable capabilities we still have for being generous to, respectively, ourselves, our kin, and socially bonded nonkin provided a mix that got those earlier humans through the Late Pleistocene, and because of continuity in gene pools, we experience very much the same mix of motivations today.

TESTING THE SELECTION-BY-REPUTATION HYPOTHESIS 11

REPUTATIONAL SELECTION

To understand why the best hunters expend great energy and take daily risks to help provision an entire band of mostly nonrelatives, Alexander's selection-by-reputation theory appears to offer considerable explanatory power. First, as Alexander suggests, the cooperative sharers are making large reputational gains as good citizens because of their beneficence, whereas the opposite holds for selfish bullies and equally selfish cheaters and thieves. Second, as I've added, free-rider suppression functions as a major selection agency. This means that seriously problematic bad guys will suffer additional fitness losses owing to potentially brutal group punishment. Keep in mind that often such active punishment comes as a reaction to a long-standing pattern of rule breaking, which means that negative reputational choices are involved as well.

Basically, the selection-by-reputation model is keyed to how people's viscera—and also their conscious calculations—respond to unusually attractive or unattractive social characteristics of others, and this can be a complicated matter when altruistic generosity is at issue. We saw with Nisa that she was actually criticized by her husband for being too generous, and when I asked my colleague Polly Wiessner about this, she told me that in general Bushmen who are overly generous in distributing food are considered to be poor partners because they are "wasting" resources. (A close analog in our own culture would be a compulsively generous big shot who squanders the family budget by repeatedly buying drinks for everyone in the bar.) On the other hand, people with really stingy social reputations are cordially disliked by the Bushmen. Thus, as far as Bushman reputations go, there can be too much altruism, which bothers partners, as well as way too little, which seriously bothers the entire group.

In analyzing the interpersonal attractions that lead to superior cooperative partnerships, we must also consider emotions like sympathy that lead to generosity, as opposed to "altruism," which as the term is being used here merely involves a measure of beneficence as this affects gene frequencies. The problem is that direct evidence for generous feelings that underlie giving is seldom provided in ethnographies. This is the case even though hunter-gatherers universally think in terms of golden rules that are designed to reinforce people's tendencies to be sympathetically responsive to the needs of others. The sympathy variable definitely is ethnographically elusive. But in my opinion it is extremely important,[1] and fortunately we have one systematic field study in which it has, in effect, been measured with its reputational effects, and also another study of marriage choices in which the social benefits of being an empathetic partner are strongly implied.

STRATEGIC RESEARCH ON THE ACHÉ

Selection-by-reputation theory was developed by Alexander with people like the Bushmen in mind, but even for the well-studied Bushmen, there's no research that focuses directly on sympathetic generosity and its social effects. Fortunately, there's one fascinating systematic investigation that does touch directly upon such outcomes, conducted among the still-egalitarian Aché foragers in South America. Their recent adaptation includes being attached to a mission and practicing some horticulture, but the Aché do continue to engage in a substantial amount of foraging. The collaborative study in question focuses on the effects of having generous reputations, which fits perfectly with both Alexander's indirect reciprocity hypothesis and with our interest here in empathetic giving. Actually, Alexander's theory isn't mentioned—possibly because the Aché anthropologists preferred to work with simpler models like kin selection, reciprocal altruism, and costly signaling. But the data presented do provide a nice test of selection-by-reputation theory, and the findings are positive.

This sophisticated study begins by focusing on two variables. One is how productive people are at subsistence, and the other is how freely they normally share their food. The object is to study how extensively others in their band will support them with food when certain hazards that are typical of a tropical forager lifestyle afflict them temporarily. These include short illnesses, bites from insects or snakes, personal injuries, and accidents, any of which can seriously affect an individual's subsistence efficiency. The odds of experiencing such afflictions are high enough that problems like these are a predictable part of Aché life.

Here's the scientific hypothesis, which I quote in full because it's so important to the theory being advanced in this book: "We propose that when temporary disability strikes individuals under conditions of no food storage, able-bodied individuals are more likely to provide food and support to those who have strong reputations for being generous and to high producers."[2]

This thoughtful investigation was based on four formalized "types":

1. *Philanthropists*, who not only are unusually generous, but are unusually productive so that overall their beneficence is extensive.

2. *Well-Meaners*, who are exceptionally generous but also are unusually unproductive, so that what they can give away is very limited in spite of their obviously prosocial intentions.
3. *Greedy Individuals*, who produce a lot but give away relatively little because they are stingy.
4. *Ne'er-Do-Wells*, who produce little and also are stingy.[3]

A sample of Aché were carefully interviewed about episodes when they were incapacitated, while their everyday sharing behavior was assessed by direct observation. As a result, it was possible to quantify how productive they were, how much food they normally gave away, and how much help they had received while unable to feed themselves adequately.

The findings show that a generous reputation definitely pays off. Unsurprisingly, the bountifully generous Philanthropists were helped the most in times of need, but interestingly the attitudinally very generous Well-Meaners came in second, even though they had so little to give away. Next came the stingy Ne'er-Do-Wells, who at least had a reason for being stingy, and last were the Greedy Individuals, who were both very stingy and also very well-off. These facts fit well with Alexander's hypothesis that people prefer to interact cooperatively with individuals with generous reputations and not with those who are known to be stingy. The findings also suggest that being sympathetically appreciative of the needs of others counts socially.

The periods of disability under study were short term. However, because food is not stored, the immediacy of Aché food procurement is such that donations by others can be important to the reproductive success of those who are disabled for even a few days.[4] Thus, all four categories of people were in need of help soon after they were incapacitated. In this context, the fitness advantages of the two empathetic types—the highly productive Philanthropists and the much less productive Well-Meaners—significantly surpassed the two selfish types in the number of sympathetic helpers who came to their aid and the amount of help given.

This interesting study shows that both the resources and the motivations involved in being generous or selfish are taken into account indigenously when contingent, "safety net" assistance is needed—and that this important temporary type of help can be given amply or in far less abundance, depending on the reputation of the person who needs assistance. Consider the fact that, even though Greedy Individuals actually give away more food to others than the much less efficient but sympathetic Well-Meaners do on an everyday basis, the latter receive more help when incapacitated. This "generosity is rewarded for its own sake" hypothesis contrasts with costly signaling or "showoff" hypotheses,[5] which basically predict that the best hunters will gain the best mating opportunities, but do not include generosity, per se, as a variable relevant to reproductive success.

The Aché pattern is instructive, for it shows that being generous can pay off in times of distress in part because of what is given to others, in terms of quantity,[6] but also in part because the emergency help is enhanced if people recognize that previous everyday assistance given to others involved giving when it hurt. This suggests that significant generosity resulting from an ability to appreciate and respond to the needs of others could have been at work in human evolution, in the context of reputational selection.

PLEISTOCENE-STYLE EVIDENCE

Ethnographic common sense tells us that LPA peoples' choices of spouses, work partners, trading partners, and band members worthy of being given extra help will be guided by the society's general values—which pointedly uphold generosity and strongly condemn stinginess. Unfortunately, to test this working hypothesis, there's no other formal study comparable to this well-focused investigation with the Aché, who do not fully qualify for the "LPA" category.

Earlier, in exploring this problem, my first move was to go to Robert Kelly's *The Foraging Spectrum*, the most comprehensive compendium on hunter-gatherer socioecological behavior put together so far. Kelly acknowledges the hazards of being a forager, but there's no specific and quantitative analysis of safety net benefits to parallel what we just reviewed with the Aché. What he does say is that "the failure to share among many hunter-gatherers in fact, results in ill feeling partly because one party fails to obtain food or gifts, but also because the failure to share sends a strong symbolic message to those left out of the division."[7]

The Aché study bears this out, for the people receiving the least help in their hour of need were the Greedy Individuals, who normally had plenty but chose to share very moderately. The Aché study's findings can also be phrased positively and with a psychological nuance. Sympathetic generosity, especially sympathetic generosity that is more costly, brings the best benefits from prosocially oriented peers. But even though Kelly acknowledges that feelings of generosity are recognized and important, basically he sees sharing patterns as ensuing from a web of obligations that leads to both requests and demands for food. This certainly is true. But what the Aché study tells us in addition is that there is a sliding scale involved and that psychologically apparent displays of generosity or stinginess are taken into account in a substantial way that can involve both positive and negative selection by reputation.

I believe this finding should hold for any LPA forager group, for the Aché continue to share basic traditional values, which favor both sharing in general and sharing with those in special need. The same golden rule thinking that inspires some individuals to be far more generous than others also influences how band members perceive these individuals, and this is part of how reputations are built.

MARRIAGE CHOICES AMONG THE HADZA OF TANZANIA

Alexander emphasizes that selection by reputation could work in supporting altruism through partnering in marriage, the assumption being that generous individuals would be given preference. Anthropologist Frank Marlowe queried eighty-five Hadza, who do qualify as LPA for-

Table VI The Traits Mentioned by Hadza as Important in a Potential Spouse†

Character	Looks	Foraging	Fidelity	Fertility	Intelligence	Youth
*Good character	Shorter	*Good hunter	Doesn't want others	Can have kids	Intelligence	Young
*Nice	Thin	*Can get food	Stays home	Once have kids	Think	
Won't hit	Good body	*Hard worker	Good reputation	Will have kids	Smart	
*Compatible	Big	*Fetch water	Likes you	Lots of kids		
*Good heart	Big breasts	*Fetch wood	Cares about home			
*Understanding	Good looks	*Will feed	Only wants you			
Gentle	Good teeth	Can walkabout				
Goes slowly	Good genitals	*Can help work				
Share words–won't fight	Good appearance	*Cook				
*Good person	Sexy					
*Good soul	Good face					
*Cares for kids						
Not bad words						
Can live together						
Won't sin						
One heart wants						
**Traits likely to involve generosity*						

†This table was adapted from Marlowe 2004.

agers, as to the features they wanted in a spouse, and he used an open-ended approach that allowed the Hadza to answer in their own terms.[8] As someone who had been working with the Hadza for decades, he then grouped the responses into categories that made ethnographic sense to him.

Hadza marriages are not formally arranged by parents, though their blessing is sought, so the principals have substantial leeway in making their choices. Table VI shows the responses as Marlowe "clustered" them, and I have asterisked responses that are likely to imply generosity as a general personal quality that the Hadza value.[9]

Here I reproduce Marlowe's findings, in which he has organized all the valid responses for choosing a mate into eight categories. The greatest number of informants referred to the Character (sixteen mentions) of a desired spouse; Looks (eleven), Foraging skills (nine), and Fidelity (six) were also salient. Under Character, nepotistic generosity was obvious in the response "Cares for kids," whereas by implication altruistic generosity seems possible in "Good character, Nice, Compatible, Good heart, Understanding, Good person, and Good soul." Under Foraging, generosity is strongly implied in all of the starred categories, which have to do with the subsistence contributions of a marital partner. Furthermore, under Fidelity generosity might have been figuring in the "Good reputation" category even though sexual reputations appear to be the focus.

The virtue of using open-ended responses is that the ethnographer doesn't impose alien categories that distort the data; the liability is that things that are very obvious to natives may go unstated. I suspect that had Marlowe given his informants a range of choices that included being generous, this would have been a frequent preference because in every hunter-gatherer ethos this is such a major virtue.[10] As it is, however, this piece of research goes far beyond most ethnographies in at least providing some major hints as to how a reputation for generosity could figure strongly in one important type of social choice.

Another African example comes from Marjorie Shostak, in a passage that characterizes how the !Kung evaluate prospective marriage partners for their daughters: "In choosing a son-in-law, parents consider age (the man should not be too much older than their daughter), marital status (an unmarried man is preferable to one already married and seeking a second wife), hunting ability, and a willingness to accept the responsibilities of family life. A cooperative, generous, and unaggressive nature is looked for, as well."[11]

Here, the ethnographic mentioning of generosity is direct, while the criteria subscribed to by the !Kung parents who are looking out for the welfare of their daughters (and themselves) correspond rather well with criteria subscribed to by Hadza of both sexes in seeking partners individually. Although my coded data for LPA foragers do not specify preferences that are salient in choosing whom to associate with in marital or other partnerships, I believe that the preferences shown in these two unrelated African groups are widespread even though many ethnographers have failed to touch upon this subject.

LESS DIRECT EVIDENCE

Making a really definitive case for selection by reputation as a major support for our human capacity for empathetic altruism will require substantial further research effort, and I would place replication of the Aché study high on any list of priorities. Unfortunately, in 2012 there are very few LPA hunter-gatherers left like the Hadza, many of whom are still actually in business as economically independent foragers.

What this preliminary analysis suggests is that ultimately there is far more to explaining hunter-gatherer sharing than a combination of selfishly tit-for-tat reciprocal altruism, resented-but-tolerated theft of meat, costly signaling by the very best hunters, or group selection effects. These models obviously are useful, but some of them seem to reduce sharing and helping behavior to pure economic self-interest, which is ethnographically counterintuitive. To fully understand human sharing and its evolution, the roles that culturally defined generosity and social reputations based on sympathetic generosity play must be brought more directly into the equation.

Meanwhile, we may start with what we know about ourselves as culturally modern humans, who, to reiterate an important and revealing pattern, often respond to televised pleas for help for needy children on the other side of the world and do this anonymously so that the motives can only be ones of generosity. We may also think about both

Thomas the Cree hunter and the small Hadza party, none of whom could conceive of not sharing with a person in serious need—even when the generosity was extremely unlikely to be repaid even on a contingent basis, because the recipient was a member of a different group. These examples may be "anecdotal," but in the absence of systematic information, anecdotes can provide useful leads.

In an important sense, with the Aché it seems to be the thought, or more properly, the feeling, that counts. This is not to say that among hunter-gatherers sharing isn't guided strongly by a sense of past obligations and past material benefits, or that selfish personal interests are not important in motivating people to engage with a system of indirect reciprocity for the "insurance benefits." My argument is that generosity based on feelings of sympathy also contributes significantly to the overall process, and that such responses are highly consistent with the deliberately, well-internalized, prosocial values that help such systems to operate as well as they do.

In this light, the Golden Rule is not just about transferring commodities from one person to another so that reciprocation will take place; it's also about fostering a spirit of generosity that can engender more generosity. And the universality of such preaching can be explained in two ways. One is that people believe in its effects as an agency that can make social relations more positive in an immediate sense. The other is that it more generally functions to oil the machinery of a system of indirect reciprocity so that the system operates more smoothly for everyone and provokes less serious conflict over the long term, which helps groups to flourish. I am not so certain that this second effect is consciously obvious to the actors involved, but hunter-gatherers do actively appreciate social harmony for its own sake.[12]

What I'm suggesting is that, even though giving to those in need on the basis of a combination of empathy and reciprocity may be difficult to measure scientifically, this is a cornerstone of human cooperation, and it's definitely not based on precise, tit-for-tat exchange. The unique Netsilik system for seal-sharing may have come rather close in combining an indirect-reciprocity-based system with a very exact structure for contingent reciprocation. But the remarkable network-based system created by the Netsilik appears to be the product of special social and ecological circumstances; their next-door cultural neighbors, the Utku, have no such arrangements.

It's worth emphasizing here that when the Netsilik cluster for months in aggregations of sixty or more, they do not try to share every seal among the entire population. Rather, they create mininetworks that link about half a dozen hunters in a variance-reduction system on the same scale as in a band of thirty, which also has about the same number of hunters. LPA sharing systems for initial meat distributions can be far less formalized; they depend on everyone's understanding and accepting the heavily contingent nature of the band-wide system in terms of who produces what—and who gets what. Giving special help to those in temporary need works similarly, but it appears to be influenced quite significantly by generous versus stingy reputations, whereas basically any group member in good standing may receive a fair share of the meat when large game is taken.

The safety-net type of sharing the Aché study investigated definitely was reputation based, and in such cases band life becomes a social stage on which good and bad reputations have been built with respect to being generous—or stingy. The game theory experiments in evolutionary economics that I have alluded to so frequently bear this out quite nicely. Once a prospective partner realizes someone is prone to be generous, a reciprocating, cooperative partnership can develop.

REPUTATIONAL SELECTION SUPPORTS ALTRUISM

If we take a close look at several dozen geographically distributed foraging societies, as has been done recently in the journal *Science*,[13] and if we focus just on the third of them that fit with our LPA category of hunter-gatherers, there's a range of band sizes and compositions that varies around some strong central tendencies that are of interest for the theories being developed in this book.

One is a tendency for just a few close kin to be coresident in the same band, which means these bands are far from being kin units. That's why I've said that kin selection models have limited applicability in explaining the band-wide cooperative systems that humans had evolved by the Late Pleistocene, unless the piggybacking "slippage" factor we talked about in Chapter 3 has been very substantial. It also is characteristic of these mobile bands for people to be moving from one band to another, and while this degree of porosity reduces the force of genetic group selection's operating with much mechanical force,[14] nonetheless I think its contributions are likely to be significant.

A further consideration is that as today some form of mostly monogamous marriage would have been universal among earlier LPA foragers. As a two-way process, selection by reputation should have tended to pair off the more generous cooperators for breeding and child-nurturance purposes, as well as for purposes of subsistence. This is strongly implicit in the findings from the Hadza study we considered a few pages back, while with the !Kung it is more explicit. Thus, all of these mechanisms—selection by reputation, reciprocal altruism, kin selection, and genetic group selection—need to be modeled simultaneously. Chapter 3 was devoted largely to studying their workings, and for the most part any of these models can operate quite independently of the others.[15]

To those models we must add kin selection, for nepotism accounts for the strong generosity that obtains within families. We must also keep in mind George Williams's thoughts about extrafamilial generosity's being similar in important ways to generosity that takes place within families. Indeed, in evolutionary terms nepotism may have served as a preadaptation for the evolutionary development of altruism.

The Hadza study hints very strongly at altruism's being a desirable feature in marital choice, but in any event people in these egalitarian bands pay close attention to whether others are sympathetic, generous with food, hardworking, or trustworthy—versus mean-spirited, stingy, lazy, or overly cunning.[16] They often are choosing their associates accordingly,[17] and this would apply to band membership, to safety nets within bands, to marriage arrangements, to routine nonmarital subsistence cooperation between families within the band, and to cooperation with people from other bands as a means of establishing long-distance safety nets or trading opportunities.[18] Unusual generosity—or a noteworthy lack thereof—can be an important factor in all of these relations.

Having a good, generous character is important, but it's far from being all important. There are a variety of purely structural constraints in choosing these same types of partners, which will include ongoing kin ties and present or past in-law relationships. But even if an individual is most likely to join a band where close kin or in-laws are living, he or she also is more likely to try to choose the relatives who are more generous or more productive, in line with character traits the Aché and !Kung were favoring, rather than relatives who are unproductive, very stingy, or both. From the standpoint of gene selection, an altruist is more likely to be more successful in cooperation with kin and nonkin alike. And when two altruists choose each other, both will be profiting in fitness terms.

From a theoretical perspective, it has been generosity based in feelings of sympathy for others and their needs that has concerned us here as a rather hard-to-study aspect of hunter-gatherer cooperation. The basis for such generosity can be sustained at the genome level if certain criteria are fulfilled, even though by definition there will be losses of fitness to be accounted for whenever generous behavior fulfills our definition of being altruistic. One criterion is that somehow the altruism must be compensated by the action of at least one of the five mechanisms discussed in Chapter 3. These include two possible piggybacking models, one-shot mutualism and long-term reciprocal altruism models, and the selection-by-reputation model. For several of these models, and also for the group selection model, free-rider suppression also has to be effective.

For LPA foragers, the compensation criterion is being fulfilled, largely, I believe, in the form of reputational benefits. Thus, in combination with free-rider suppression my candidate as a main contributing mechanism to explain the limited but well-exemplified and socially significant altruism of hunter-gatherers has been selection by reputation. There's a reason for giving these two social selection mechanisms such priority. This lies in the fact that reputation-based social selection is likely to gain special power precisely because the selection process is based on choice. This makes it similar to Darwinian sexual selection, but with respect to its efficacy among hunter-gatherers, selection by reputation might actually be going sexual selection one better.

INTERACTIVE EFFECTS

It would be useful if someone adept at evolutionary mathematics could model all the mechanisms we have discussed in combination, weighting each model for its probable contribution to the maintenance of altruistic genes in human gene pools over the past 45,000 years and more. However, the working hypothesis I'm offering here is that the selection-by-reputation model is likely to be especially powerful because not only is this type of social selection driven by well-focused, consistent preferences, but it also involves a two-way choice process. When the Hadza or the !Kung follow their preferences in choosing partners to mate with, or for that matter just partners to work with, one major effect of their choice behavior will be that generous and fair cooperators will be in a position to successfully choose others who are similarly altruistic and that on both sides fitness will profit.[19]

If the sympathetic altruists (along with hard workers and trustworthy people) are tending to pair up with others who are similarly socially desirable like themselves,[20] this leaves their less desirable counterparts tending to pair up with one another—and suffering the disadvantages of less effective cooperation. Obviously, the advantage goes to the pairs of worthier individuals who are reaping the fruit of superior collaboration—and at the same time are unlikely to suffer from free riding.

This is significantly different from the pattern of having drab peahens who on a one-way basis choose resplendent males whose superior genes bring females more reproductive success. It's as though both male and female fowl were developing attractive tail feathers to advertise their fitness—and the males were choosing the females with the best tail feathers at the same time that they themselves were being chosen on the same criterion. Were this the case, the power of sexual selection, which is widely acknowledged to be remarkably strong, would be stronger still because both parties would be simultaneously displaying superior fitness—and choosing it.[21]

When geneticist Ronald Fisher became fascinated with Darwin's treatment of sexual selection, and with the fact that such choice-driven selection can result in "exaggerated" traits, he thought that costly peacock tail feathers might be explained in terms of "runaway selection," which results from the escalating interaction of astute female preferences with male signs of genetic superiority.[22] When both the traits being chosen and the choosers are involved on both sides of the equation, I believe this could further intensify such effects. Mathematical modeling would be useful in testing this hypothesis, but as a cultural anthropologist obviously that is not my forté.

Richard D. Alexander differentiates between "sexual selection" and "reciprocity selection" (I've been using the more descriptive term "selection by reputation" for the latter, for the sake of clarity for a general audience) as he considers this two-way element in the choice behavior that is inherent in selection by reputation:

> Sexual selection is a distinctive kind of runaway selection because joint production of offspring by the interacting pair causes the process to accelerate. . . . The defining feature of runaway selection is not acceleration, however, but the tendency of the process to go . . . much further beyond adaptiveness . . . than is ordinarily the case in the myriad compromises among the conflicting adaptive traits that create and maintain the unified organism.
>
> This aspect of runaway selection may hold for reciprocity selection, in which, unlike in sexual selection, both parties can

> carry tendencies not only to choose extremes but also to display extremes. In social selection . . . an individual can play both roles, of chooser and chosen.[23]

Of course, one thing that Alexander means when he says that traits selected by reciprocity selection can go "significantly beyond adaptiveness" is that costly altruistic traits might be supported by this type of selection, as well as any other traits that appear otherwise to be maladaptive for individuals but are positively involved with reputational choices that are mutual. This may be the best one hypothesis so far to help explain human altruism—but only, I believe, if it is combined with the very effective kind of free-rider suppression I've been emphasizing throughout this book.

WHY FREE RIDERS HAVEN'T SIMPLY VANISHED

Trivers was correct when he suggested that moralistically aggressive groups could come down hard on detectable cheaters. However, I've suggested that with earlier humans, and still earlier with Ancestral *Pan*, it was selfish bullies who were taking the main free rides by competitive actions that favored their own genes at the same time that the genes of their less selfish and less powerful victims were being disadvantaged.

In Chapter 4 we saw that in fifty LPA societies these bullies appear to have been executed far more often than thieves or cheaters, or, most likely, than sexual offenders, which means that their selfish aggressions were seriously problematic for their groups—and also that these powerful free riders were paying a steep price in terms of loss of fitness. We also saw that when earlier humans were intent on being egalitarian, physically punitive types of social selection probably would have been acting far more powerfully than today because initially there was no conscience to help restrain the bullies.

Thus, the earlier stages of free-rider suppression were likely to have been harsh indeed, and we may reasonably assume that the more selfishly driven alpha types, those prone to take risks, were bringing severe and quite frequent punishment on themselves until a conscience evolved to aid them in controlling themselves and thereby in avoiding such dire consequences. As bullying (or cheating) free-rider types increasingly became able to restrain themselves from actions that would bring on punishment, they would have suffered fewer and fewer reproductive disadvantages, so with the existence of more effective self-control there's no reason to believe that their genes should have gone out of business entirely or have even come close to doing so.

Keep in mind that even though LPA bands are so highly egalitarian, this does not mean there's an absence of competition. Males compete for hunting prowess, and males (and females) compete for mates. Indeed, egalitarianism itself is based on competition between a few stronger individuals and the subordinates who unite to oppose them. Thus, if a person can channel his or her competitive tendencies in directions that are socially acceptable, and at the same time curb their expression when they will make for fitness-reducing punishment, selfishly competitive tendencies can be quite useful to fitness.

Refraining strategically from aggressive behavior is well exemplified by Inuttiaq, for his efficient, self-conscious evolutionary conscience put him intuitively in touch with his personal social dilemma, and it enabled him to strategically inhibit many of his aggressive responses rather than expressing them in ways that would make his fellow Utku seriously apprehensive—and perhaps prone to dire action.

Not all men prone to despotism manage to curb their despotic tendencies so efficiently. In fact, when Scandinavian explorers first contacted one Eskimo group in Greenland, they noted that a very dominant shaman had killed serially and was being treated with great respect by his group.[24] They left before the dominator was killed, so that was not witnessed, but we may be quite confident that somehow he was disposed of. We've vicariously witnessed a similar !Kung execution in the Kalahari, and other Inuit groups do away with such men.

In Chapter 7 Table IV showed us that such executions are quite widespread. But whether would-be despots successfully inhibit themselves or not, their actual potency as free riders who can take serious long-term advantage of altruists and others is very limited. Either like Inuttiaq they're afraid to take the free ride, or like /Twi the aggressive !Kung Bushman they'll be killed for taking their aggressive free ride too actively.

This leads to an important theoretical point that requires some further emphasis. George Williams's mathematical portrayal of free riders and altruists assumed that free riders were designed (by evolutionary process) to exploit altruists and thereby disadvantage their genes.[25] As a result, altruistic genes could never reach fixation in the gene pool concerned. And if new altruistic genes were to appear as mutants, free-rider mutants would soon appear to drive them out of business.

When we bring in the conscience as a highly sophisticated means of channeling behavioral tendencies so that they are expressed efficiently in terms of fitness, this scenario changes radically. Over time, human individuals with strong free-riding tendencies—but who exercised really efficient self-control—would not have lost fitness because these predatory tendencies were so well inhibited. And if they expressed their aggression in socially acceptable ways, this in fact would have aided their fitness.

That's why I believe that both free-riding genes and altruistic genes could have remained well represented and coexisting in the same gene pool. Genes that made for bullying free riding could have been useful because they were providing a useful competitive drive, whereas genes that made for altruism could have been useful because altruism was being compensated by reputational benefits and by other compensatory mechanisms we have discussed.

This useful self-inhibition wasn't perfect. In today's bands, we still have the occasional active bullies and cheats, who are prone to take major free rides because their own reckless optimism about what they can get away with leads them astray,[26] because their dominance or deception is compulsive, or because the feedback their consciences provide them with is faulty. We've also seen that it was bullying that caused more of the social problems, and that it did so more frequently. Let's reconsider the numbers. In Chapter 4, Table I, bullying ("Intimidation of group") dominated the reported executions, and in Chapter 7, Table III, the most frequently reported acts of deviant social predation were acts of domination by intimidators (with 461 cites for the ten societies), whereas acts of cheating were mentioned only 42 times and for only half of the ten societies sampled even though the central tendency was quite noteworthy. Thus, even though cheating free riders have been academically center stage ever since Trivers made them so famous in 1971, for humans it appears that the major free rides have been taken by politically forceful selfish dominators, whose victims include not only generous altruists who equal them in power but are much less selfish, but also anyone else who is less inclined or less able to dominate others.

Aggressively selfish behavior continues to be a widespread problem today among LPA foragers, and it's obvious that individually variable, selfishly aggressive tendencies are still with us. If the people so endowed were free to express these propensities without social inhibition, then fair-minded, generosity-driven systems of indirect reciprocity simply would not work. With respect to human cooperation, it's fortunate that these selfishly aggressive propensities have been susceptible to a remarkable degree of control from within and also from without. From within the human psyche an evolutionary conscience provided the needed self-restraint, while externally it was group sanctioning that largely took care of the dominators and cheaters who couldn't or would't control themselves.

AN EVOLUTIONARY SEQUENCE

At this point I'd like to set forth a historical sequence, starting at the ancestral beginning, as a way of summarizing much of what has been said in the preceding pages—and as a way of keeping my promise to make the natural history of moral origins more historical. To start with, most likely primitive "altruism quotients" in Ancestral *Pan* were as modest as they seem to be in today's bonobos and chimpanzees. However, in comparing their degree of innate sympathetic altruism with ours, we must keep in mind that in groups these apes have no way to amplify their cooperation through golden rules.

Our behavioral reconstruction also tells us that a noteworthy if rudimentary potential for (nonmoralistic) group social control existed in this ancestor, a potential that was being directed exclusively at bullies as a resented and readily identifiable type of selfish, competitive, exploitative free rider. This means that the selfish behaviors of certain aggressive individuals could be curbed, even though basically ancestral social orders remained quite hierarchical. This ancestor did have a significant capacity for self-recognition, but in the absence of an evolutionary conscience, its self-control was based just in fear of retaliation, and a capacity for submission.

We cannot be sure how abruptly the next phase of moral evolution began, but it could have involved the escalation of similarly antihierarchical social control to a point that rather than profiting, stronger individuals more often were paying a significant price for their attempts at opportunistic, free-riding domination. This could have led to genetic selection in favor of an enhanced capacity for self-control that involved something new: although fear of retaliation by subordinate coalitions continued to slow these bullies down, as a protoconscience developed my hypothesis is that rules could now be internalized as well and that this led to a more sensitive adjustment of individual behavior to group preferences.

It's impossible to estimate when the capacity to identify emotionally with rules began to seriously affect overall patterns of social behavior and, hence, selection outcomes. *Homo erectus* with its relatively large brain is at least conceivable as a possibility. However, the position I've taken is that such social selection had to become quite decisive starting a quarter of a million years ago, when a still larger-brained archaic *Homo sapiens*—toward the end of its career—began to hunt large, hooved mammals and depend on their meat. For groups of people in bands to have been really efficient at this, it's very likely that meat-sharing had to have been well equalized, and this efficiency could have been crucial for group or regional survival when climate change made local environments challenging. As we've seen, another possible effect of decisive sanctioning would have been that disempowered alphas would have had problems in reproductively controlling a band's females. This may well have opened the way for monogamous pair bonding to develop, or to develop further.

I've taken 250,000 BP as the magic number for the likely beginning of really strong conscience evolution, but new facts could make for an adjustment of this hypothesis. If it were discovered that some archaics were depending on intensive hunting 400,000 years ago, then that might be our date instead of 250,000 BP. The same would be true if some new *Homo erectus* site showed systematic hunting of large ungulates a million years ago, although the brains involved would have been much smaller. However, these hypotheses would be difficult to reconcile with the fact that archaic *Homo sapiens* went from multiple butchers to a single butcher between 400,000 BP and 200,000 BP.

None of this chronology can be definite, but my sequential theory is that the first stage of moral evolution resulted in an evolutionary conscience, and once we became moral, two new patterns were able to develop. One was selection by reputation that favored altruists, and the other was a moralized version of free-rider suppression, which would have targeted not only bullies, but also thieves and cheaters. At that point, we may hypothesize that altruists were beginning to pair up assortatively with other altruists. And with the help of an evolving conscience, more sophisticated strategies of social control would have enabled people to reform those deviants who were more responsive to group rules and to group wishes, rather than injuring or killing them or banishing them.

Unless new evidence is found, by the time people became culturally modern our moral life was basically complete, as LPA hunter-gatherers know it today and for that matter as we ourselves know it today. We had both a sense of virtue and a sense of shameful culpritude, and we understood the importance of human generosity well enough to promulgate our predictable golden rules across the face of a then thinly populated planet. We were a people who in important ways had conquered our own abundant selfishness—even though that conquest required constant vigilance, and considerable active tweaking of the types we have spoken of.

The Evolution of Morals 12

WHAT, EXACTLY, ARE "MORAL ORIGINS"?

As we've seen, it's all too human to be curious about beginnings. Indeed, it's likely that somehow our brains were set up to think that way, for all humans seem to think about how we became so different from other animals in the important matter of morals. The intuitive philosopher in all of us just naturally wonders about how things got started, as opposed to assuming that somehow they must have been eternally in existence, and the answers human beings have come up with in the moral sphere are various, to say the least, as well as fascinating and often colorful.

For those who are theologically inclined, questions of moral origins may conjure up Adams and Eves who contemplate forbidden but inviting knowledgeable fruit hanging from readily accessible low branches in trees. Or there may be images of this same pair, fallen and shamefully hiding their newly private pudenda behind the plucked leaves of a nearby ficus, as seen in many Renaissance paintings. Long before such stories were written down, however, the entire world was dealing in oral traditions that perpetuated similar myths to satisfy similar human curiosities.

Anthropologists can assure us that today in virtually any ethnographically described nonliterate society, people will be thinking deeply about questions of origins—be they of the physical world, of people, or of morals—and that they'll be encouraging rhetorically gifted specialists to tell them their origin stories from memory. The unusually imaginative and detailed myths of the Navajos testify to this, and just in the single *Ichaa* tale we heard not only about the earlier development of humans from mothlike entities, but about moral origins with respect to the incest taboo and how it came into being. That story was told by a gifted Navajo mythmaker named Slim Curly in the 1930s.

Naturally, this widespread human concern with origins is found among our LPA foragers, so it's safe to say that this "around-the-campfire," mythological approach extends far backward in time. The Garden of Eden story, Darwin's personal interest in moral origins, and the very writing of this book would appear to have been *culturally* preadapted for us in the African New Stone Age, 45,000 years ago and more. The same ever-curious, problem-solving minds that make us ask exactly the same questions today have driven certain authors—including a number of the scientific writers we're about to discuss—to write popular books having to do with moral origins.

Those earlier, oral-tradition mythologies suggest that first people—and subsequently their morals—were simply created out of whole cloth. The same reasoning is found in the formal religious belief systems that eventually followed—hence we have Adam and Eve as probably two of the more striking mythological figures ever to have emerged on an all-at-once basis. In contrast, the theory of natural selection brings us to a rather different kind of interpretation, for in comparison biological evolution is a process that builds gradualistically on its previous achievements—even when it's punctuated.

The fact that gene mutations continually come into being is at least *suggestive* of unprecedented novelty, but mutant genes and random genetic drift are simply nature's way of providing fodder for an overall Darwinian process that basically moves through time completely blindly and quite gradually, building on its own past precedents. For a biologist like Ernst Mayr, this ongoing process is by definition both dynamic and continuous at the same time.[1] Nonetheless, when our common sense tells us an evolutionary event is both novel and important, we do tend to talk about origins. This was the case with Darwin, when in 1859 he entitled his first book *On The Origin of Species*. This remarkable naturalist obviously thought deeply about moral origins as well, even though his own high standards of scientific reasoning didn't permit him to come up with even a tentative historical sequence in *The Descent of Man*.

Thus, "moral origins" is a venerable item in our scientific vocabulary, but we must keep in mind that the origins involved will have been based on preadaptations, and the mythmaker's "whole cloth" idea must be set aside. Mutant genes surely were important, but basically nature has always liked to combine older building blocks with newer ones and then combine this product with still newer ones. That provides the continuity that Mayr spoke of.

I've suggested that moral origins took place gradually, over thousands of generations, through natural selection that gave us a conscience, including a sense of shame. Just like other selection events, this involved not only preadaptations but also, in all probability, significant environmental changes. My hypothesis has been that the immediate agency that created a shameful conscience was punitive social selection, so in fact there could have been two environments that helped to shape moral origins, depending on which of the three chronological hypotheses proposed in Chapter 6 was operative.

More at a distance was the changeable natural environment, which provided delicious, nutritious large ungulates to hunt, along with materials for fashioning some serious hunting weapons, plant foods to gather, water to drink, means of shelter, probably some curative herbs, and occasional periods of stress. It was the *social environment,* however, that provided the more immediate selection forces, and this social niche was in part created by humans themselves.[2] The original, punitive type of social selection gave us a conscience, but by providing such efficient free-rider suppression, it later made it possible for altruistic traits to evolve as strongly as they have.

WHEN DID MORALS ORIGINATE?

With regard to explaining moral origins in a concrete way, philosopher Mary Midgley half a decade ago took a relatively pessimistic view in *The Ethical Primate:*

> We can indeed wonder how and when our remote ancestors did actually come to be troubled with a conscience, how they became aware that they could make free choices, how they developed moral concerns to the extent that every human society now has them. But we are unlikely ever to have more than the faintest, most tantalizing indications about this strange process, indications which can mislead us as easily as they can help us. They are misleading not just because they are scanty, but because of our own remoteness. Even if we could somehow listen in at some crucial point and had help with the language—or proto-language—the situation would be so unimaginably strange to us that we would stand little chance of grasping it. So we have here a gap which we have to fill in, like other historical gaps, as best we can from indirect evidence, from what comes before and after, and from careful comparison with other species.[3]

I'm probably a bit more sanguine about making an evolutionary reconstruction of moral origins than Midgley is, but then I have spent the last decade immersed in data on chimpanzees, bonobos, and LPA hunter-gatherers, trying to make many of the kinds of comparisons and connections she calls for. I also have recognized the barriers she identifies, and for that reason I have not attempted to describe the protoconscience beyond acknowledging its existence.

At the same time, I've been watching carefully to see what archaeologists were coming up with. Two studies by Mary Stiner that we encountered earlier have been critical to the more specific theories espoused in this book, and they were published within that same ten-year time frame. They told us that humans became dedicated large-game hunters a quarter of a million years ago and that their mode of butchery also changed in ways that could have been socially significant. It has been my view that if these various types of indirect evidence could be joined in a synthesis, and if relative plausibility was the scientific standard used to test the overall hypothesis, at least a substantial beginning could be made toward a fuller understanding of how we and we alone became moral.

I have tied moral origins to the major political transition of earlier humans from being a species that lived hierarchically to becoming one that became devoutly egalitarian. The theory I've proposed can be stated simply: what put this very decisive brand of egalitarianism so firmly in place was the ability of politically unified groups to "outlaw" and punish resented alpha-male behavior. The impact was profound, for this put an evolutionary premium on self-control and also began to suppress free riders in ways that were all but uniquely human.

What I cannot specify with current evidence is whether the humans who were just beginning to go after large ungulates were perhaps still nearly as hierarchical as their ape ancestor had been—or whether the transition to egalitarianism was already well under way. Aside from subordinate males wanting more personal autonomy or improved access to females for breeding, an additional factor, earlier on, might have been a desire to share more efficiently whatever carcasses were acquired and butchered, which included not only some megafauna but also, surely, smaller game like that hunted by Ancestral *Pan*.[4]

Whatever the earlier scenario, I have held that this overall political transition to egalitarianism could have been significantly accelerated, and made definitive and culturally institutionalized, at the point when humans were becoming active-pursuit hunters. Because variance reduction was so important to their nutrition, at that point they were obliged to live in bands with a fair number of other hunters, and somehow they had to efficiently share out the sizable but not enormous game they were killing, because as entire hunting teams they were routinely investing so much energy in its pursuit. These are hypotheses that have, I think, some specific empirical support—and also some general plausibility. But that is for the reader to judge.

We might also try to key such egalitarian developments to the crude archaeological evidence we have about brain size, for it's at least logical that the larger the brain, the more that autonomy-loving subordinates could have been capable of effectively ganging up to improve their competitive position against high-ranking dominators in gaining meat—or females. But there's no way at all of telling when brains became powerful enough socially to permit the creation of a decisive and stable egalitarian order.

I've suggested that archaic *Homo sapiens* might conceivably have already been *fully* egalitarian before a quarter of a million years ago, when intensive hunting began. If that were the case, we could turn things around and theorize that definitive egalitarianism paved the way for hunting, rather than vice versa. For moral origins, this is not important; what matters is that the strong social control that made earlier egalitarian orders possible led to the evolution of our human conscience.

What I feel most confident in hypothesizing is that from a quarter of a million years forward archaic humans in bands, with their obvious needs for efficient meat distribution, had a great deal to gain by being engaged in the same intensive, effective, *general* suppression of alpha-male behavior practiced today. This would have brought with it some really aggressive free-rider suppression, and in all probability it would have involved not only the many individuals who had stronger alpha tendencies, but also those with other antisocial proclivities like cheating or theft, which also would have seriously interfered with effectively equalized meat-sharing. Any such behavior would have aroused people's ire in a multifamily band that was intent on sharing its favorite nourishment and was facing periodic but not calamitous scarcities that made such sharing very important.

At the beginning of what might be called the egalitarian transition, fear of receiving aggression would have been the primary and primitive (ancestral) psychological mechanism that drove the natural dominators to submit when their groups opposed them. And most likely the physical conflicts between resentful groups and their less restrained dominators would have been far more frequent than today, driving social selection strongly in favor of a conscience.

This conscience would have evolved as self-control based on rule internalization became more efficient, and evolving a degree of conscience that Mary Midgley or you or I would recognize surely involved changes to the brain. This process was likely to have taken at least 1,000 generations, depending on how strong the social selection was in its early and presumptively often quite brutal operation. That's only 25,000 years, and probably a more reasonable figure would be 2,000 to 4,000 generations (50,000–100,000 years), but as we've seen, up to 8,000 generations could have been available with any of our three scenarios.

At the time that something close to a modern conscience would have evolved, and this would have included a shameful blushing response as well as emotionally identifying with rules, we may definitely speak in terms of moral origins. The point of comparison is the submissive, fear-based self-control of a domesticated dog or wolf or of a bonobo or chimpanzee. Once we'd acquired a full moral sense, which included thinking in terms of socially attractive virtue as well as shameful vice, gossiping accordingly about our fellows, and having a sense of our moral selves, the difference had become profound.

The following hypothesis is consistent with any of the three tentative sequences developed earlier in the book. First, it was angry, punitive social selection by groups that first gave us this physically evolved conscience. Second; it did so by making free-riding bullies and others who couldn't control their antisocial impulses pay genetically for their "crimes," and afterward similar forces, now endowed with morality, continued to vigorously suppress the behavior of would-be free riders—

which made it much easier for altruistic traits to evolve genetically. This second phase of moral evolution might have begun at latest in the vicinity of 200,000 BP, but this is a guess. And because altruists not only were protected against free riding but also were pairing off with other altruists, the selection forces that genetically favored altruistic traits could have been powerful enough to require only a few thousand additional generations to make us fully as altruistic as we are today. Thus, by around 150,000 BP, when we may at least hypothesize that anatomically modern African humans were on the verge of cultural modernity, we could have been well on the way to becoming moral beings and to becoming significantly more altruistic than our more distant ancestors had been.

HOW "TRUE" IS THIS SCENARIO OVERALL?

That is the story of moral origins as I am able to piece it together with present information, and it differs radically from any that has been told before. Perhaps in rendering this account I have allowed myself to become rather venturesome as a scientist, done so because the questions being addressed are so humanly important and also so intrinsically fascinating. But that is difficult to say because the many-faceted, holistic scenario I have developed here is not easily tested on a scientific basis that includes clear-cut "falsification."

Of course, many of the component hypotheses are readily subjected to scrutiny, such as the parsimony-based reconstruction of ancestral behaviors or the reconstruction of culturally modern behaviors in human foragers as of 45,000 years ago. There are many other areas where alternative approaches are possible, and these include the very definition of morality, the focus on shame, and the rather expansive definition I have used of the evolutionary conscience. However, if we consider the moral origins theory advanced in this book as a unified whole, the best way to test it is simply to judge its overall plausibility in comparison with other theories of moral evolution.[5]

MORAL ORIGINS THEORIES SINCE DARWIN

The scientific territory covered by moral origins is wide, just as it should be for such a large and relatively unexplored topic. However, my interest here does not extend to evolutionary ethics, as discussed by sociologist Herbert Spencer, Thomas Huxley, and others more recently.[6] What the purview of this book does include is the mechanisms that have been active in the evolutionary development of shame, virtue, extrafamilial generosity, and moralistic group social control, and the task I set myself has been to write a pointedly *historical* natural history of moral origins, with full attention to details that include what preceded these origins and what happened to human social life afterward.

One interesting twist on the moral origins theme arrived over a century ago with Friedrich Nietzsche's *On the Genealogy of Morals*.[7] This well-known writer followed Darwin in providing a philosopher's version of moral origins with a strong evolutionary flavor, and his origins scenario was specific and, like mine, quite political, if rather fanciful. But basically the argument was more about questions of power, turn-the-other-cheek weakness, and anti-Christianity than about how morals came into being. In a sense, the power theme does link Nietzsche's work with what I have done here, but the beauty of hunter-gatherer egalitarianism is that the weak, in joining forces to control the strong, themselves become powerful.

As an archaeological treatise James Breasted's *The Dawn of Conscience* sounds extremely promising in its title,[8] but the idea seems to have been that to gain a purchase on moral evolution, we must turn merely to the ancient Egyptians. Darwin would not have agreed with this, nor do I. However, I do believe that Darwin would have heartily approved of Finnish sociologist Edward Westermarck's monumentally documented *The Origin and Development of the Moral Ideas*,[9] which was published before Breasted's work and just a quarter of a century after Darwin's death.

Westermarck's remarkable analysis makes ample use of the then available nonliterate ethnography, and it covers some of the topics I have concentrated upon here, such as moral emotions (including altruism), conscience, and capital punishment. It does so insightfully, but in spite of its overall brilliance, today this interesting work is known mainly for Westermarck's almost offhand hypothesis about incest,[10] which was mentioned in a previous chapter. This powerful synthesis deserves better acknowledgment on many other scores, largely as a precursor for today's evolutionary psychology with its focus on emotions. However, Westermarck did not follow Darwin's general approach by bringing a strong historical dimension into his evolutionary analysis, and with the information available to him at the time this would have been fairly difficult.

My guiding principle has been that the historical dimension is critical to a full explanation of moral origins, and that it was simply because Darwin hadn't the needed information that he was obliged to suggest a "byproduct" type of argument with respect to the conscience. Today, a number of other scholars are actively exploring the moral origins question mainly from the ahistorical perspective of "adaptive design," which also is taken from Darwin. But I find it curious that in spite of our vastly improved knowledge from archaeology, and in spite of our growing capacity to make reliable ancestral behavioral reconstructions, this historical dimension has been set aside to the degree it has.

Perhaps part of the answer is that in grappling with the question of conscience, Darwin set a precedent, and, as loyal Darwinists, scientists who respect his work have simply assumed that in such matters a historical analysis is out of the question. But another piece of the puzzle is that modern academicians tend to put themselves into specialist compartments, whereas Darwin's curiosity knew no disciplinary boundaries.

Edward O. Wilson did end his classic interdisciplinary work, *Sociobiology,* with a tentative historical analysis of human social evolution that included matters of morals in the sense that altruism was a focus.[11] However, a few years later in his *On Human Nature,* which dealt much more directly with morality, he did not pursue this historical dimension further.[12] I believe it was this second work of Wilson's that set the standard for well-known popularizations like those of Matt Ridley, Robert Wright, and James Q. Wilson, along with Michael Shermer.

Ridley's popular scientific work *The Origins of Virtue* is essentially a sociobiological tract in that it sticks to models like kin selection and reciprocal altruism.[13] Like Robert Wright's *The Moral Animal,*[14] Ridley's book essentially lacks any in-depth historical dimension. The same is true of James Q. Wilson's *The Moral Sense,*[15] which is written more humanistically but, like more technical academic works in evolutionary psychology, is equally ahistorical if it is compared with Darwin's way of writing natural history. A similar but theoretically more far-ranging work is Michael Shermer's *The Science of Good and Evil,*[16] which breaks with this sociobiological tradition in one way: it gives some serious consideration to group selection theory as this has been espoused by Ernst Mayr and David Sloan Wilson.[17] But it, too, is essentially ahistorical. Also essentially ahistorical is Marc Hauser's *Moral Minds,*[18] which basically takes a linguistic approach to moral origins.

In a more technical volume edited by philosopher Leonard Katz, which is entitled *Evolutionary Origins of Morality,*[19] four long chapters provide a nice sampling of the scientific diversity currently encountered in this field. I have already mentioned the first essay, by Jessica Flack and primatologist Frans de Waal. They talk about empathy as a major *building block* for moral evolution;[20] here, I've used the more technical term "preadaptation" to the same effect. Their building block approach ties in nicely with a historical evolutionary approach, and it has been as a result of reading their work that I have emphasized human sympathy (they refer to it as empathy) so heavily in the preceding pages.

In Katz's book my own anthropological chapter comes next,[21] and it deals with the prehistoric role of social sanctioning and conflict resolution in the natural selection of moral behavior, with some hints of the free-rider suppression theory I've developed here. In the third chapter philosopher Elliott Sober and biologist David Sloan Wilson continue the arguments they made in *Unto Others,*[22] endeavoring to establish group selection as an important factor in moral evolution and to expand its theoretical scope.[23] (In their important book *Unto Others* there's a great deal of evolutionary theory, and some excellent use of ethnography, but again not very much emphasis on historical process, per se.) The final chapter, by evolutionary philosopher Brian Skyrms,[24] is heavily involved with relevant mathematical modeling, and characteristically it is ahistorical and oriented to explaining morals in terms of game theory and adaptive design.[25]

HUMAN NATURE MATTERS

The great majority of the contemporary work on the human nature aspect of moral evolution is more in line with this last approach of Skyrms. To test such models, most often the primary data are generated in laboratories, usually with children or college students as subjects, and the findings are tested against criteria of evolutionary design, a mode of analysis that (as I've said) does stem directly from Darwin. A large number of evolutionary psychologists and evolutionary economists, including Ernst Fehr in Zurich, do this type of work, and in the field of morals some of the overall flavor of evolutionary psychology is exemplified by the title of an article by Dennis Krebs: "The Evolution of Moral Dispositions in the Human Species."[26] But design, not holistic natural history, is the approach throughout.

Among a growing coterie of evolutionary economists, elaborations of game theory of the type that originally inspired Robert Trivers have been used to investigate morally relevant behaviors like human generosity, our sense of fairness, the uses of punishment, and the punishment of nonpunishers.[27] In addition, Robert Frank, notably in *Passions Within Reason,* has contributed significantly to our evolutionary understanding of conscience and moral emotions.[28]

An interesting recent debate that ties in with the egalitarian theory I espouse has to do with whether, when people in these experiments go out of their way to punish those who make "unfair" offers, they are motivated by a spiteful or otherwise resentful need to retaliate or whether they are expressing an *aversion to inequality*.[29] In subsequent experiments, it appears that by the age of seven to eight, children act on feelings in this latter direction—which helps to build the case that antihierarchical feelings are an important and evolved component of human nature.

Evolutionary economists such as Sam Bowles and Herb Gintis have explored the impact of social control through "strong reciprocity."[30] And with an emphasis on fighting between bands, Bowles has also explored the possibility that group selection might have worked robustly to support altruistic traits in the Late Pleistocene;[31] he has suggested that prehistoric warfare, in combination with major genetic differences between different prehistoric forager groups, could have generated significant forces in favor of group selection.[32] If we factor in the moralistic free-rider suppression that I have been emphasizing so heavily, this may provide a major, multilevel formula for explaining the evolution of altruistic traits.

HISTORY MATTERS

Contemporary archaeologists and paleoanthropologists have done a remarkable job of explaining historically the physical evolution of humans and their material culture over time, taking their basic methodological historical cues directly from Darwin. They also have studied the cognitive side of prehistoric cultural evolution, to good effect.[33] When it comes to accounting for the moral side of our evolution, however, these and the other scholars I've just discussed have not readily adopted the historical approach that Darwin himself would have preferred to use.

To many, my concern with writing the natural history of morals more historically may seem rather old-fashioned, or even quixotic, but my aim has been to provide as complete a scenario as possible for moral origins and to do so by employing the rich, holistic type of evolutionary analysis that Darwin used to such good effect—whenever his data allowed him to do so. I could have wished for still better data, but I have provided a number of hypotheses that, I believe, may be useful to future explorations of moral origins in a number of fields, as better data do become available. If some of the present working hypotheses eventually are modified or even replaced by theories that seem more plausible, so be it.

IS HUMAN EUSOCIALITY UNIQUE?

The preceding chapters have made it abundantly clear that moral origins involved some radical changes in our behavioral potential. Yet our ape ancestors, in spite of their lack of feelings of shame, at least had the potential to impose "rules"—as individuals but also as groups—even as they responded to rules imposed by others. It's our sense of virtuous good and shameful evil, along with our universal and symbolically stated love of altruistic generosity, that sets us apart.

If my treatment of moral origins has focused heavily upon homologous continuities, there has also been what appears to be some outright "novelty." The symbolic language that allows us to gossip in such specific terms is one such advance. Blushing with shame is also a major evolutionary anomaly in this sense.[34] Could such blushing be a signal, sent to others, that somehow redounds to the fitness of the signaler? Or could it have evolved as a way of signaling to oneself that social danger is being courted? I hope that someday we'll be in a position to make some better educated guesses.

Both socially responsive shameful feelings and the accompanying bodily flushing, which is intimately involved with self-awareness, seem to be unique to humans. The same is true of having a deeply felt sense of right and wrong coupled with a capacity to internalize group rules of conduct—a capacity that is based in personal feelings of being morally worthy or unworthy. Our human degrees of altruism and cooperation may not make us unique among living beings, but what other animal gets there the way we do—by knowing shame or by developing a sense of virtue? And what other animal deliberately amplifies its own altruism because it understands cooperative societal functions well enough to do so?

With respect to the longstanding altruism paradox, our *moralized* type of extrafamilial generosity certainly is new among mammals, including the fascinating ones we've discussed that live in *kin-based* eusocial colonies that also are held in place by group selection.[35] Analogically, the hypercooperative naked mole rat in fact can easily outdo our self-sacrificial generosity as these rodents achieve an antlike social organization. But they do this through underlying mechanisms that appear to be very different from ours, and clearly they do so in the absence of morality. Obviously, this applies also to the social insects, such as many species of ants, bees, or wasps.[36] They, too, may outdo us in contributing "selflessly" to their societies, but if we look for homologies, we are sorely disappointed: basically, they are doing this in the absence of anything like an internalizing conscience, gossip, group social pressure, shameful blushing, or moralistic capital punishment by angry, morally outraged bands.

Yet it certainly is true that when we think about how far the human potential for cooperation can develop on the ground, it's a beehive that readily comes to mind. We have only to think of Egyptian or other pyramid builders—or of rural Hutterite communities or hippie communes or, for that matter, the Nazi Wehrmacht's most dedicated (and ruthless) elite corps—and the parallels are there.[37] Nonetheless, the social insects merely provide a striking analogy, which shows that natural selection can stumble into the making of a collectively oriented species in more ways than one.

If we compare any of these eusocial colony dwellers with Ancestral *Pan,* our ape ancestor's group-level cooperation pales by comparison because it was limited mainly to ganging up against conspecific bullies or collectively threatening neighbors. Yet it was this socially limited ape ancestor that had the useful building blocks in place as far as moral origins and the evolution of a human style of cooperation were concerned.

As a far less cooperative antecedent, Ancestral *Pan* does offer us the following homologies. As reconstructed, this species possessed capacities not only for self-awareness, perspective taking, and dominance and subordination, but also for formation of antihierarchical, counterdominant coalitions. In addition, mothers were empathetically socializing their offspring and providing them with cultural learning models.[38] This was a remarkable and fortuitous array of preadaptations, and as building blocks all of them, I think, have been important or even crucial to our moral evolution.

Yet these ancestral apes cooperated as entire groups mainly in routinely guarding their territories, in sometimes putting down alphas, and possibly in mobbing predators.[39] Like Ancestral *Pan*, today's chimpanzees and bonobos are never likely to build pyramids or organize themselves to distribute meat in a fair and basically equalized manner, in spite of all the ancestral precedents the three of us have shared. And a profound evolutionary question that I introduced earlier still lingers. These two *Pan* species have had exactly the same amount of time that we have (about 6 million years) to develop a conscience like ours—or at least to develop some kind of shame-based feelings of right and wrong. Why have only we managed to do so—in spite of all these shared primitive characteristics? If the analysis in Chapter 5 was correct, and I admit that I wasn't bending over backward to give living apes the benefit of the doubt, they haven't even come close. And overall the analysis in this book suggests that this might be because their group social control has remained so limited that fear-based self-control responses have been able to do the job satisfactorily.

THE IMPACT OF SOPHISTICATED INTENTIONS

If we consider the three fundamental (and competing) "interests" that our genetic nature is designed to serve,[40] I've emphasized that basically they weigh in heavily in favor of egoism and then, after egoism, nepotism. Both individual self-interest and family interests are straightforwardly selected aspects of our genetic potential,[41] and unarguably they're strong. I've emphasized this repeatedly because it's so basic, and I don't believe there's an evolutionary biologist, anywhere, who would say no to this position. And then there's our still rather mysterious "altruistic quotient," which ever since Darwin, and especially over the past half a century, has been clamoring for further explanation.

Even though at present scientists are far from being able to identify any functionally very specific human behavior genes, which includes any that might be involved with extrafamilial generosity, I've suggested—noncontroversially, I believe—that overall this inherent propensity in favor of being generous outside the family has to be rather weak by comparison. Yet as Sober and Wilson's work suggests, in everyday contexts this rather modest altruistic potential can be greatly culturally amplified in its expression—if it is actively and purposefully reinforced by social communities that believe in things like social harmony and the Golden Rule.[42]

I must emphasize, further, that when such phenotypic amplification is accomplished on an insightful basis, this intentional input tends to "focus" genetic selection processes in certain directions—notably, altruists who uplift social life are consistently favored while stingy individuals and those who behave disruptively are regularly disfavored.[43] These intention-bearing inputs are made possible by our large brains, and in a sense the intentionality involved does bring a certain *purposive* element into the process of social selection—and therefore natural selection. Our promotion of generosity helps altruistic genes to be selected reputationally at the same time that our punishment of uninhibited, free-riding bullies or cheaters works against the selfishly aggressive genes that they carry. The advantage goes to the altruists as long as they stay politically united, and therefore to their genes as they are represented in the gene pool.

The human preferences that orient both punitive and prosocially oriented social selection are on a relatively long genetic leash,[44] and because they involve flexible choices among alternatives, their effects can be highly facultative, going in quite different directions. For instance, in the face of starvation, only a human being can frame the resulting social dilemma in terms of life, death, and continuation of the family and then consciously choose among the perceived alternatives. And only a group of humans can talk over the dilemma of whether to try and reform a seriously entrenched bully, as opposed to quietly asking his brother to step in and do away with him. Intentions, combined with high intelligence, do make a difference.

I am putting these last ideas forward even though biological scientists normally don't appreciate the use of any language that smacks of "teleology" in conjunction with evolutionary process. In their book, and mine, natural selection by definition has to be basically blind and not in any way "guided." However, even though when hunter-gatherers' preferences affect gene pools this is wholly unintentional, when they go about shaping and implementing their own immediate social policies, often they know exactly what they're doing. The fact that LPA foragers so predictably use symbols to try to amplify altruism is a testimonial to this, and as a result selection-by-reputation helps to shape gene pools. By itself, this one potent human social preference provides any evolutionist with some serious food for thought.

Both our human cultural capacity in general and many of the specific things that we're genetically prepared to learn very early in life—such as acquiring a language or helping others in need or having an aversion to inequality—give us an unusual capacity to shape our societies in certain directions, and not in others, and to do so consistently over evolutionary time. If we look at the everyday cultural priorities and vocally manipulative behavior of LPA foraging people in their bands, they have managed to give substantial weight to tweaking altruistic tendencies and to ensuring cooperative outcomes. Innately helpful tendencies are reinforced both in child-rearing and later on, and I'm convinced that LPA foragers have a reasonably good grasp of what they're doing when they reinforce them.

If culture is so intimately entwined with biology, how can we effectively factor out the purposive element in culture to ask what its effects may be on gene pools? The hunter-gatherers in my sample of fifty societies do this kind of tweaking quite regularly by insisting that meat be divided according to the rules, by actively encouraging generosity, by cracking down severely on major free riders, and by trying very hard to manage conflicts before they explode. They often do this preemptively by discouraging the self-aggrandizing or deceptive aspects of their fellows' behaviors when these are likely to result in victimization—or conflict.

This strategic use of punitive social selection improves the reproductive success of everyone save for antisocially selfish types, such as overly dominant or deceptive deviants. And I've held that there's a zero-sum game going on: the greedy deviant's loss, for example when he's stopped from hogging the meat, is everyone else's gain. Thus, it's good for everyone's genes to be part of a group-wide coalition that sees to it that a few alpha males are prevented from monopolizing a modest-sized carcass like an antelope or zebra and sees to it that all share in the proceeds more or less equally.

An important theoretical point is that such culturally based purposeful inputs are both part of natural selection and a product thereof. Thus, their effects have gone beyond shaping everyday group life prosocially, for *they have helped to shape our gene pools in prosocial directions that are similar*. I believe that these powerful brains of ours have been making all of this possible for thousands of generations, and one major and totally unintentional side effect has been the conscience that originally made us a moral species. Another has been our unusual propensity to practice generous behavior outside the family, which evolved through a variety of mechanisms and which, I propose, humans have deliberately amplified in order to facilitate better cooperation.

THE ADJUSTABILITY OF GENEROSITY

Tendencies to extrafamilial generosity just naturally come into conflict with innate selfishness and nepotism—even as they're being amplified by the impressive array of cultural practices we've met with, and even as social selection makes them useful to fitness. For scores or hundreds of millennia, the balance among these three in their everyday expression has made possible the cooperation we are famous for, and if we look at hunter-gatherers' overall meat-sharing patterns, our well-amplified, innately generous impulses can be seen as oiling the wheels of a collaborative hunting-and-sharing system that has been highly useful to the interests of, respectively, individuals, families, and bands as a whole.

From the standpoint of reproductive success, this cooperative system has worked superbly in what I've called normal times, even in bands like those of the ever-contentious Bushmen, or of certain Australian Aborigines we've mentioned, with their routinized grousing about not getting fair shares of meat. The bottom line is that all LPA hunter-gatherers do share large carcasses immediately after they kill them, that they do so in similar ways, and that they seldom get into *serious* conflicts over this meat even though it's so precious. This is their normal mode of operation, and even when the making of hostile demands is pronounced, this simply produces social pressures that help to keep the system of sharing working for everyone.

This is the cooperation our forbears are so famous for, and it takes place in normal times, when the meat supply is adequate—or at least is not woefully inadequate. But I've presented an untold part of the human cooperation story that is equally important. For our species, all times haven't been normal—far from it. We've seen that in situations of serious scarcity tendencies to extrafamilial generosity will begin to lose out and that even nepotistic helpfulness can be set aside.

Such flexibility stood our species in excellent stead when attempts to stay with normal, "good times" modes of altruistic, contingent meat sharing could have led to total extinction of a hungry, embattled regional population had it continued to live—and share—in a good-times mode. Once a band's equalized sharing system was abandoned, and sharing declined to the level of nepotism, this might have permitted at least a few lucky or unusually adept families to survive by cooperatively subsisting on their own until better conditions arrived, or until migration to a different region with better possibilities could be accomplished. In still harsher situations, as we've seen, a similar argument can even be made with respect to individuals acting just on the basis of egoism.

When Leibig's Law comes into play sharply, and the already-limited quality of human generosity becomes seriously strained, on average the individuals who can adjust their responses accordingly will gain a fitness advantage. Thus, it's adaptive to fit in well with a band or family unit that cooperates well—but it's also adaptive to strategically step back and become more selfish if that's the only way out. This certainly could have been adaptive, but obviously it was very stressful emotionally for the moral beings who in earlier hard times were keeping our species in business for us.

I believe it's when the chips are down, and true famine strikes, that the tripartite division of labor within human nature we've been discussing can be evaluated most accurately. And what this tells us is that when food supplies are adequate, cooperation can pay off quite handsomely and that the rather modest dose of innately prepared extrafamilial generosity we possess can, in fact, go a long way because of shrewd cultural reinforcement—and because generosity can beget more generosity.

The result has been an efficient—but fragile—capacity for cooperation that has had the virtue of being quite flexible. That's why today we can cooperate effectively not only as hunting bands but also as tribes or as chiefdoms—or as individual nations. Whether this will work for today's global community of nations may be a different matter, and I shall briefly address that problem in the Epilogue.

A THOUGHT EXPERIMENT

Just because humans have been capable of making prosocial choices that shape their societies and affect their very gene pools, does this mean that our species has evolved as far as it could in this direction? I doubt it. Consider, first, what might have transpired had the Late Pleistocene Epoch continued for another 50,000 years instead of beginning to seriously phase out about 12,000 years ago, and had the invention of agriculture never even occurred. Possibly, this additional 2,000 generations of exclusive (and often highly precarious) hunting and gathering might have made us evolve genetically so that our social systems would work more smoothly than they can today, or so that our moral responses would be somewhat different, or so that our modest but important gift of generosity-based altruism might have become stronger—or even could have dwindled, though this seems unlikely.

Thus, today's innate moral capacity may amount to a mere evolutionary work in progress—even though in so many respects it seems to have been doing its main evolutionary job of contributing to individual reproductive success and getting us through times when the ecological chips were down. If in fact our moral potential is still changing, this may not fit very well with our self-concept as *the* moral animal, which of course has some self-congratulatory "finished-product" philosophical undertones. But my job as an evolutionary anthropologist is to tell it like it is, and we can only guess about this.

If conscience evolution began fairly abruptly in terms of an "egalitarian revolution" starting just 250,000 years ago, things may still be evolving in this respect. If it began much more gradually, long, long ago, with an authority-hating, autonomy-loving *Homo erectus* or early archaic *Homo sapiens* that perhaps wanted better access to females, our moral evolution is likely to have become more genetically stabilized—the idea being that at 45,000 BP an equilibrium would have been reached with respect to an optimized conscience and an optimized "altruism quotient." We may never find a way of assessing all of this, but it's interesting to think about.

The scientific study of Pleistocene bottlenecks probably is not yet complete,[45] in that additional findings may be on their way. But we may well have just barely made it in terms of navigating some really dangerous junctures when relatively small numbers of human survivors hung on in refuge areas while their regional colleagues perished by the hundreds of thousands.[46] Our evolving flexible moral capacity may have been an important part of this picture, and if we assume that these close calls did occur once in a while, it makes sense that even with the facultative adjustments we were able to make, these flexible social capacities were barely up to the job of keeping us alive at the species level.

One obvious implication is that another 50,000 years of Pleistocene instability might eventually have done us in, pure and simple, through sheer bad luck—at some time of cruel adversity when even an "every man for himself" response couldn't keep just a few of us in business. But, conversely, had these wildly varying selection pressures continued, they might have allowed us time to develop some still better (or different) mechanisms for coping. Just how this might have affected our moral capacity is difficult to say, but the selection forces involved certainly were likely to have favored moral flexibility.

Basically, the evolutionary "destiny" of our species has been up to chance—unless you believe that a micromanaging Divine hand was protectively overseeing the process. I don't. I believe devoutly in dumb luck as far as the basics of biological evolution are concerned, and this affords little comfort if as a human being, you'd prefer to feel watched over and "special." Of course, I've argued that intelligent intentions on our own part may have helped us genetically to become by nature moral, but we certainly never *intended* this to take place. Nor did we design ourselves to make it through the Pleistocene, even though I believe we were smart enough to continually fine-tune our contingency-based meat-sharing systems in times of ecological adequacy, and smart enough to abandon them when the chips were really down.

All of which is to say that we cannot take our present moral potential or our present capacity to cope with environmental crises to be either finished products or outcomes of biologically enlightened intentions. We arrived at our moral nature in the great evolutionary arena of chance, even though our very immediate purposeful inputs influenced the process in directions that were prosocial. Indeed, if we set aside this hypothetically extended Pleistocene as an evolutionary fantasy, and consider the actual Holocene present we live in, our moral capacity surely continues to evolve at the level of genes, for some key environmental constraints have changed, and for urban humans surely they are changing further as I write this.

FUTURE HYPOTHESES ON MORAL EVOLUTION

I've hypothesized that 45,000 years ago culturally modern band-level societies were continuing to help shape our human gene pool as people favored individuals with good reputations and punished active free riders, just as their forbears had done. The social context was a cooperative band with usually four to seven families, a band far too egalitarian to let any one person run the show. Today's huge, highly organized, socially and politically hierarchical modern societies differ in the ways that problems of social deviance arise, and in how they are handled by highly centralized systems of law and order, so as I said, this evolutionary story hasn't necessarily ended. In fact, its trajectory is likely to be changing.

As just one instance, in spite of police detectives and computerized data banks, sociopaths too often go undetected in our large, anonymous societies, and their frightening genetic footprint could, in theory, be increasing. We need only to think of a modern serial rapist, who in an intimate hunting band would know that he could be easily identified—and that as a practical matter he'd better avoid expressing such behavior or he'd soon be killed. Today, if undetected, in theory he can gain a major genetic advantage even though in many parts of the United States abortion interventions would tend to reduce this. And over a few thousand generations, such advantages might be adding up—even though, very fortunately, at least a portion of these conscienceless monsters are taken out of circulation so that they can't stock our gene pools. In many other ways, presently unfathomable, we may assume that our evolutionary course could be gradually changing because, in these modern settings, the selection scenarios have changed.

We do now have a genomic baseline for future studies, even though at present we can't identify a single very specific "moral gene." And a few generations in the future, we may have identified some of the genetic mechanisms that help us to behave egoistically, nepotistically, and altruistically, along with others that make for sympathetic generosity, domination and submission, and a variety of other socially significant behaviors that are relevant to morality, including our shame responses. Perhaps we may even be able to understand the basis for social dispositions that make us prone to blush and find some clues as to how they might have evolved. This may seem optimistic, but then it would have taken an optimist in 1950 to predict that the double helix code would soon be cracked.

At some future point we could also compare a future genome with the one we possess now, but unless that future were very distant, this probably would tell us very little—unless gene selection can proceed far more quickly than we believe it to. This is not out of the question socially if we consider all of the types and levels of selection that are likely. These include two-way runaway social selection, kin selection, piggybacking that involves mistaken kinship, reciprocal altruism and mutualism, group selection, and Simon's docility selection. All could be contributing to the stabilization and further evolution of traits that make us generous to kin and nonkin alike and that make us moral.

We could also compare current genomes with prehistoric ones, where the necessary DNA is retrievable for *Homo erectus,* for archaic *Homo sapiens,* and for anatomically modern humans. Some of these opportunities are already available, and with all this new information we might be able to ask some new kinds of questions, or answer ones I've been struggling with here, such as fixing a point in time at which a conscience began to evolve or providing clues as to exactly how or when we were likely to have acquired blushing as a sign of moral distress. What we need is a new Watson and Crick, who can open up the study of social behavior genes in spite of their surely very high degree of complexity.

New substantive findings could lead to novel ways of framing evolutionary questions, and many of the hypotheses explored in this book could become more directly testable or even falsifiable in the sense that philosopher of science Karl Popper famously used the term.[47] But I must emphasize that Popper eventually decided that the testing of evolutionary theories constituted a special case, as far as rigorous falsification methods were concerned. Basically, scientific rules for the "building evolutionary scenarios game" have to be more permissive, and that is why I have referred so often to relative plausibility. It is also the reason that I have dared to set forth tentative but specific scenarios like the "large-game hunting necessitated egalitarianism" one.

For the present, the methods and information I have relied upon must suffice, and the various theories I have generated here must be judged largely by how much sense they make as working hypotheses, as well as by how well they fit into the larger pictures I have built. Such evaluation is not easy, and some scientists may prefer to throw up their hands—or even suggest that I may be "hand-waving" my way into one big just-so story. However, I believe the question of moral origins is important enough to merit investing the effort and taking a few chances in terms of setting up working hypotheses that, in combination, may stimulate better working hypotheses for the future.

My credo has been that such scenarios, even if partly wrong, can lead to more satisfying scientific explanations in the future. And I do believe that the Darwinian evolutionary scenario I have laid out here to be an advance over Darwin's own story, which in 1871—all of 140 years ago—took the conscience to be an apparently inevitable side effect of our human intelligence and empathy, and looked solely to group selection as the way to explain extrafamilial generosity.

Our consciences and their functions are, indeed, tied to intelligence and sympathy, but the message of this book is that the conscience has actually evolved on its own in ways that can now be hypothesized on the basis of relative plausibility. Such a theory can be built not only by considering prehistoric changes in the natural environment, but also by giving some serious weight to prosocial human choices as these have shaped our social environments, sometimes quite brutally, and in turn have helped to shape our gene pools.

As I was writing this final chapter, I found myself wondering how convincing the scientific answers I have proposed in the preceding pages will be with the passage of yet another 140 years. I can only hope that my story here, like Darwin's, will be seen as being incomplete but far from wrong–and worthy of continuation.

Epilogue: Humanity's Moral Future

This book has addressed an ancient curiosity about human origins, and about our moral origins in particular, and as such it qualifies as pure science. Here, briefly and as an afterthought, I shall carry the evolutionary analysis into an at least partly knowable future as I consider the more immediate practical prospects for our world moral community of nations.

We won't be considering the future of the human gene pool here, but rather how what is essentially a Pleistocene human nature impinges on our civilized moral future. In times to come, the *cultural* side of our ongoing moral life faces at least one profound challenge. Rather than rising from the vagaries of an often perilously unstable Pleistocene environment, this challenge is partly one of our own making. It has to do with the further development of our entire planet as one big moral community, that is, as a Durkheimian type of society that can promote cooperation and regulate deviant behavior.

Over the past 12,000 years we humans have increased the sizes of our social communities from bands to agricultural tribes to chiefdoms to nations, and at all of these levels we have basically succeeded, and succeeded fairly well, in sorting things out so that destructive internal conflict doesn't rule excessively and bring us all down. When such chaos phases in today, we call the nation in question a state that has failed. But most nations are far from being failed states, for they function quite well with their formal legal systems and their institutional approaches to law and order—institutions that basically are thousands of years old and can be traced back to Hammurabi's laws in Mesopotamia, inscribed on a black stone tablet, and beyond.

In several important, functional senses a powerfully centralized nation is still quite similar to the self-regulating, band-type of moral communities we've discussed at such length. This is the case even though small bands sharply dislike anything smacking of centralized authority, whereas people in nations know that they need some authority in order to avoid serious internal conflict and possibly civil war.

Like nations, bands are highly aversive to conflict; indeed, a dislike of angry tensions and social disruption is an important part of being human. However, the means of conflict management vary. Bands, like nations, rely heavily on social pressure and mediation, but ultimately bands can rely on avoidance at a distance—either the band can split up or else one party to the conflict simply moves to a different locale and the conflict's over. *Fini.* With nations things are quite different. Conflicting factions within nations obviously cannot move in space, which is why ultimately national stabilities have to be based on centralized coercive power—power sufficient to hold down socially disruptive deviance and, if necessary, step in and crush an incipient internal conflict. Bands, in contrast, can make do quite nicely with persuasion, mediation, and long-distance avoidance.

When it comes to the entire culturally diverse global society of nations that inhabits this planet, we have yet to find any really *effective* means of international social control and conflict resolution.[1] Indeed, at any given time the number of smaller wars, between and within nations, is shocking. Worse still, there's always the possibility of a really major war, which by today's standards means a nuclear conflict that can affect the health and livelihood of every nation and every person in the world. Thus, when it comes to our global community of nations, in this sense we're hanging in a major political limbo.

In practical terms, what we've done is to carefully design a world government that *looks* like an effective national government, but then we've made sure that it can be sabotaged from within when it comes to being decisive in really important matters of war and peace. I have in mind, obviously, the effective powerlessness of the United Nations General Assembly, and the Great Powers' absolute vetoes in the Security Council. In an important sense, then, we've designed a world moral system that too often, and especially with any serious differences of opinion that involve Great Powers, lacks the teeth to sanction deviant nations or to intervene in conflicts when these become dangerous to our entire planet.

With this less than potent "world government," we face profound global political challenges. These range from local genocides and conventional wars to the specter of worsening nuclear proliferation, to the threat of terrorist use of bacteriological or chemical or even nuclear weapons, to future problems we cannot even predict. And we continue to face the distinct possibility that our planet could be sullied radioactively by nuclear accidents still worse than those we've now experienced, or be all but destroyed as a usable habitat by outright nuclear warfare between nations so equipped. Those are just the obvious problems.

Unlike an LPA foraging band, our world of nations is far from being economically egalitarian. No sage prognosticated, as the Cold War ended, that a Medieval-Crusades type of global conflict with potentially nuclear terrorism as its objective would so quickly replace it, even though both sets of tensions seem to have been based importantly on resentments of *have-nots* toward *haves*. We must assume that future sources of envy-based conflict could be both equally unexpected and still more insidious. But what we can count on for certain is a future collision of superpowers as China builds its economic and military strength and the United States quite possibly declines. Thus, we may face another simmering "cold war" that could become susceptible to its own unpredictable dynamics in the absence of effective international control, with similarly enormous, overkill nuclear arsenals on both sides.

The future of global morality—and global social control—will have to be watched with care, for as the means of inflicting grievous damage on others become more sophisticated, more varied, more readily available, and more widespread, and as an ever-divided "community of nations" continues to be far less than potent in containing many of its clearly apparent threats, our world system of law and order seems increasingly precarious. This is the case even though humanity's statistical rate of killing in warfare has been radically abating since 1945.[2] One obvious problem is that—reminiscently of hunter-gatherer bands—this huge world community of nations resolutely resists forming any efficient type of a supergovernment, an empowered one that could do for a world of nations what a single successful nation's centralized government does for its people.[3]

Our world is probably too large, too diverse, and too dangerous to continue to conduct its affairs informally in the hopeful style of a well-united egalitarian band that can readily coalesce as a moral community when needed, do so usually without people taking sides, and resort to avoidance if need be. As basic problems, globally we face the sheer scale and number of the political units involved and the profound cultural differences among some of our nations. And then there's a basic problem that stems directly from our human political nature: like the individual hunters in a band, these nations are too intent on their sovereignty to allow one big supernation to be built, with a trustworthy central world government strong enough to ensure the rule of international law and guarantee the peace—and if need be, do so invasively.

The last thing we need is a well-armed world of nations behaving like one big failed state, but the potential at least looms. We do have the 1949 Geneva Convention's "humane" rules of (conventional) warfare, so fortunately our global community is not entirely without laws—even though it resists creating the all-encompassing, institutionalized centralized command and control needed to back them up. Two things we do have to work with, however, are our evolved and shared sense of morality and the fact that most of us, as nations, agree on certain matters, such as the nature of basic human rights, the undesirability of poverty and disease, and the need for self-determination.

Another note of hope is that we form a species that has evolved over hundreds of thousands of years to understand its own societies and help them to function—by suppressing antisocial deviance, by socially rewarding those who are altruistically generous, and by otherwise "tweaking" our social systems to make them work better for us. The foregoing chapters have made this abundantly clear. In addition, in our bands and also later in our more complex societies, we have created national safety nets that "insure" us against personal disaster, and on occasion, at least, have behaved as unified moral majorities when we felt threatened by lawless, greedy, socially disruptive deviants like Hitler or Saddam Hussein when he invaded Kuwait. We've also tried hard to manage our internal conflicts and contain their destructiveness, and the nations of this world do try to mediate the conflicts of others.

All of these important and valuable features of prehistoric and contemporary human life are nascent in a seriously underfunded, deliberately disempowered United Nations that can only be in a position to do these things effectively and consistently if the five Nuclear Great Powers happen to agree—and also are willing to cough up the money. We must hope that somehow, eventually, these well-evolved potentialities will be expressed far more effectively in helping to shape a stable world society of nations, many of them nuclear, that has to be our future greater moral community.

Perhaps one past lesson for present and future major powers to consider would be that not only applications of power but also extensions of generosity can be a strategy for social success, and that sometimes a generous approach, in spite of the risks, may pay off handsomely in the long run. After World War II, for instance, America's popular and massive Marshall Plan was a generous (and politically useful) move that helped to build a successful and nonconflictive

Europe, while more or less altruistic foreign aid to other parts of the world was also forthcoming on a substantial basis. The United States was viewed as a rich and generous nation that deserved the goodwill of others, comparable to the unusually productive and also unusually generous Aché individuals we met with, who were helped abundantly in their own time of need. However, in this present century America has had neither the will nor the budget to follow this precedent, and our reputational standing is at low ebb because economic generosity has been so strongly eclipsed by one maladroit political power play—the second invasion of Iraq—which in the world's eyes violated another nation's sovereignty and was not based on a consensus by a world moral majority. This was comparable to a bully's throwing his weight around in an egalitarian hunting band, and America's global presence changed for the worse.

In bands, everybody shares the same culture, and in an important sense their generosity-based cooperative systems of indirect reciprocity can be based on a sense of personal trust that underlies any potential actions of the group as a moral community. Furthermore, people are egalitarian, which means that their political and economic pie is divided more or less equally. For a huge, competitive community of ethnocentric and religiously divided nations that includes enormous (and too often growing) differences between haves and have-nots, to accomplish something similar involves far greater challenges.

If like a foraging band our global community of nations is vehemently unwilling to trust a centralized system of command and control, we must hope that *somehow* these challenges can be met with the same degree of insight and realistic goodwill exhibited by our Pleistocene hunting forbears, when they realized that if they wanted to live well as hunters they had to merge their competing interests and cooperate, and they proceeded to do just that in the absence of strong chiefs. That is one note of hope. Another basic reason for optimism lies in our evolutionary gifts of sympathy and generosity. Such feelings apply very readily to our children and to other kin, also to our friends and socially familiar neighbors. It is also within our potential for them to apply to others much more distant with whom we feel culturally bonded, and sometimes, at least, to total strangers.

In a very general and very important way, I've suggested that in human forager communities, generosity and the resulting altruistic acts grease the wheels of cooperation. However, in today's world community the further our potential cooperation extends from home, the more its tenor becomes unpredictable. The world's nations may be reaching out to unknown victims of a natural disaster one day, and the next day a member of this same "community" may be committing genocide against its neighbor, quietly sponsoring vicious attacks on others by "guerrillas" whom the victims designate as terrorists, or covertly or openly trying to bring down a government it doesn't like. It's because our potential for sympathy and altruism is relatively modest, that it can be eclipsed so readily by less prosocial psychological dispositions that have come down to us from Ancestral *Pan*.

Where we moderns are similar to our hunter-gatherer forbears is in our vulnerability to conflict. A band is vulnerable because it will not allow sufficient centralized power to develop to make possible some usefully "authoritative" means of conflict resolution to go to work, especially when its more powerful members come into conflict. Exactly the same is true of our global community of nations. In combination, the Security Council veto and a noisy but politically impotent General Assembly see to that. If smaller nations are considered along with large ones, at any given time the number of ongoing wars in our worldwide "community" is still staggering, and if you add in the threat of a serious nuclear miscalculation our planet becomes a truly dangerous place. Unfortunately, after over half a century of this dire threat, we're getting used to the risk. And familiarity breeds inaction.

In coping with disruptive behavior a foraging band can be quite active. It readily coalesces into a moral community because it is bioculturally evolved to be moral, and living in moral communities provides a very immediate forum for gossiping individuals to agree on the threats they face and to cope with the deviants in their midst.

As their fear of a dangerous deviant mounts, they can increase social distance, and in the case of a really serious general threat, they can agree very privately (and very decisively) to do away with him—if avoidance or ostracism or banishment can't do the job.

Our world of nations makes partial attempts in a similar direction, for manipulative formal boycotts and once in a while active blockades are attempted with a rogue nation's reform as the goal. However, cross-cutting alliances often make effective international ostracism difficult, and sterner steps are very difficult to agree upon. Furthermore, with their vetoes the five nuclear Great Powers are selfishly exempt—and unfortunately they also are prone to back their allies, including ones that are nuclear.

Bands can solve a really serious social problem by killing the deviant. Anything analogous to capital punishment at the level of a nation is out of the question—unless a dictator can be taken down. But too often one nation's dictator is another nation's useful ally, and in any event the United Nations is not in the business of nation building. The basic problem is that any real and universal centralized power with the capacity to punish would be taken as a potential threat to all, and the main sovereign bullies on this political stage (that same Nuclear Club that was formed in the midtwentieth century) have both the most to lose politically and their vetoes to ensure that they don't lose it. In a number of ways, then, individuals in foraging bands have far more to work with in curbing or stopping deviance, especially serious deviance, than do the culturally and religiously diverse nations of the world community we live in today.

People in a band have some truly major advantages. They share the same culture. They also speak the same language, and they know one another personally. They gossip together, which builds trust. And they know that often they can leave their band, if necessary in a hurry, if a local situation becomes too conflictive. A world of culturally disparate nations, from which there is no exit, is a different kettle of fish. Nations are fixed in space, so if they can't settle their differences in any other way, they have to fight wars.

There does exist something called world public opinion. All of our nations' foreign ministers understand this well, for they play to it constantly. In that particular sense our community of nations is, in fact, very much like an LPA forager moral community writ large. In the days of the Cold War, some grand theater played out with the United Nations General Assembly as a venue for the "United States versus the Soviet Union Show." However, there was no way that the rest of the nations—the ones being imperiled by the two alpha superpowers—could use public opinion to seriously rein them in. This international moral stage continues to serve us today from time to time, and it is something that could be built upon. But even in the absence of such a formal international forum, world moral opinion will coalesce simply because, as a species, we are moral—and because television exists.

A great question for our global future is whether we will rise to the occasion as threats to our entire world community become more complicated, still less predictable, and possibly far more perilous. Very scary at first, the straightforward bilateral nuclear balance of power we lived with for the second half of the twentieth century became simply another fact of life. This was partly because the players on both sides had large populations, and enormous infrastructures to lose, were the tensions to get out of hand. Even so, history has taught us that the Cuban missile crisis was in fact a true crisis, in which the leaders of two nations, in spite of being moral, were playing a game of chicken that could have brought untold destruction to parties outside the conflict. Today, as a picture in our minds, that outcome remains difficult to conjure up realistically, because of its sheer horror.

An already dangerous balance of nuclear power became still more dangerous with the entry of India, Pakistan, and North Korea into the nuclear arena. To complicate matters, at this writing Iran seems to be arming against a nuclear Israel. Add all of this up, and it seems that a general situation of potential peril and distrust among nations may become too exacerbated for any more effective world system of governance to be gradually built—unless some event (short of total catastrophe) serves as a warning and galvanizes the world's nations into action.

In fact, and grimly, a possible catalyst for progress in world governance, and for world security, might actually be a *limited* global disaster. Imagine, for instance, a smaller nuclear war that would seriously poison our planetary atmosphere but leave most of our world population and economy intact. If a calamitous World War II taught the Europeans not to fight, perhaps a small (and equally destructive) nuclear war might galvanize our entire world of nations to create a similar and safer union.

This is indeed a bleak prognostication. But realistically, with national sovereignty being as sacred as it is, it may be the best we may hope for if a safer world community is to be built in the face of nuclear proliferation. Meanwhile, free trade at least makes us more interdependent in the economic sphere, and, as we've seen, economic interdependence was a major factor in LPA hunter-gatherers forming efficient moral communities that regulated the use of large game. In this sense, as a social catalyst modern free trade may be functionally analogous to meat-sharing among our forager predecessors. People who depend heavily on one another in the basics of making a living are likely to become more efficient at resolving conflicts—and to learn that a trusting type of generosity can pay off quite handsomely in situations where mutual assistance based on indirect reciprocity is useful to all.

Also on the positive side, we all share a basic, prosocially-oriented moral capacity, which is inherent in being human. This at least enables us to reach out to distant others when safety nets are needed, and, we can hope, it helps world political leaders—those with normal consciences—to think twice before they start brutally hurtful wars. The underlying sense of sympathy for other human beings will always serve as a counterweight to conflict, and in a community that feels itself to be "one," as we've seen these prosocial feelings can be systematically and effectively amplified for the greater good, precisely because people do have consciences.

That's how we worked things in the bands that basically got us through the Pleistocene, and these dynamics have continued in force as we've moved on to live in farming tribes, then chiefdoms and kingdoms, early civilizations, and modern nations. Now, finally, we live in a world community of nations that is at best a work in progress—and at worst could become a failed state writ very large. From bands to nations all people develop similarly moral communities in several important respects. For instance, judgmental public opinion exists at all of these sociopolitical levels; attempts are made to manage conflicts; informal rules or laws are agreed upon; and criminal behavior is deemed to be punishable.

From the viewpoint of an evolutionary psychologist, Steven Pinker suggests that our rates of killing people in warfare have been declining quite drastically for some time, even as the use of nuclear weapons has become morally taboo.[4] Some of this peace-bonus effect surely comes from a simple fear that wars with nuclear weapons will be unwinnable for either side, but the moral component is very important, and it does exist prominently in a world moral community that remains only tentatively structured to do its job.

Attempts to build a (totally powerless) League of Nations almost a century ago tell us that even in the 1920s the need for *some* kind of morally based command and control at a global center was obvious enough. Today, with a somewhat more potent United Nations, we seem to have exchanged the reality of recurrent and very costly but survivable large-scale conventional warfare for the risk of *total* catastrophe. Thus, at least in an actuarial sense, today the stakes are even higher.

What, then, are our global possibilities insofar as they may be "predictable" in certain ways from our evolutionary past? One guess would be that a benign and generous superpower might dominate the entire world and impose political order without dominating too strongly. America enjoyed this possibility after the Soviet Union collapsed. However, the second invasion of Iraq, as just one of many sovereign nations ruled by seriously offensive dictators, didn't help us to stay in this morally acceptable, altruistic role. By this act, an aggressive Bush conservative elite, with the compliance of a sheeplike Democratic Congress, spent not only our treasure but also our political and moral capital for many years to come.

Interestingly, one reason this ongoing political enterprise was so costly financially was that America, once the idea of permanent (anti-Iranian) military bases was eschewed, felt a moral obligation to create a stable nation in Iraq before it would be conscionable to withdraw. Thus, as a nation we've been condemned for an invasion that basically went against the global community's mores with respect to national sovereignty, but we have received little credit for at least staying an expensive course with respect to nation building once Iraq was seriously "broken."

If the United States for a time did have a flourishing benign role right after World War II, soon the Cold War saw the generosity image fade away as the sponsoring of military alliances and warfare by proxy became global preoccupations, and major conventional warfare returned with an ugly vengeance not only on the Korean Peninsula but also in Vietnam. With respect to being helpful to other nations, American foreign policy continues to be dominated by a highly self-interested political emphasis as we dedicate the great bulk of our foreign aid to shoring up internationally controversial regimes that are prone to keep major world conflicts alive. Too often, this dilutes our idealistic messages in promoting democracy and makes us widely unpopular. In terms of world public opinion, most likely we have politically disfigured ourselves by failing to look at the big picture with its important moral intangibles, and we may well have relied too much on our power and far little on our generosity as a nation.

As an evolutionary anthropologist who tries to view things from a distance, I keep noticing a fundamental political problem. The LPA bands that evolved our genes for us were fiercely egalitarian, and the price they paid for this political equality was having to do without the benefits of centralized command and control that could have kept their bands from occasionally splitting asunder when a social problem or conflict did get out of hand. In a hunting band, it's the militant sovereignty of the individual hunters that decentralizes politics and keeps things that way. In a global community, it's love of national sovereignty that does exactly the same thing. However, people in a band are basically economic equals, whereas our world of nations is very far from being egalitarian in this way. This economic inequality can be seen as a special engine that helps to drive international conflict, and it stands in the way of creating a more effective international order.

The frightening balance of power among competing, economically unequal nuclear nations at least may have ended very large wars of a conventional type. But probable dangers inherent in this morally reinforced balance of terror, which is subject to both technological and human error, must be multiplied by the number of nations with nuclear arsenals, and our world order is susceptible to coming seriously unglued if ethnocentric hatreds get out of hand. Both India and Pakistan and, in the near future, Iran and Israel, come to mind, and in some cases the combined nuclear arsenals, if employed, would be large enough to very seriously threaten our entire planetary environment. With such heated conflicts, the "taboo" on nuclear attacks could be set aside, just as it could have been in the Cuban missile crisis.

In this very political world of ours, continued nuclear proliferation seems likely not only because the nation in question's military potential will be enhanced, but also because basic political "respect" on the world stage all but skyrockets with a nuclear capability. It's worth noting the major hypocrisy when five nuclear nations that are already enjoying such respect preach against proliferation, and thereby try to deny to lesser nations the right to improve their international status.

Power balancing is another ancestral trait, as described for chimpanzees by Richard Wrangham.[5] In this context, any really *serious* world domination by any one nation does seem unlikely. This is because the would-be dominator would still be vulnerable to devastating attack—as long as antinuclear defense systems remain as fallible as they seem to be. Thus, among the major nuclear powers the political dynamics are similar to those in a politically and economically egalitarian band of well-armed hunters, who are obliged to respect one another's lethal fighting weapons—and the possibility of ambush—even though some of the hunters may be much stronger than others.[6] In a real sense, nuclear weapons have created a kind of modern *political* egalitarianism that is similar, but this applies just to the nations in an erratically expanding Nuclear Club.

If the world were inflicted with a mercifully *limited* nuclear disaster, expectably the shocked surviving nations might fearfully set aside their differences, compromise their respective autonomies, and at least take some further steps in the direction of creating a safer world order. Logically they would use an orderly, politically centralized, multiethnic nation as their model, and already in place is a United Nations General Assembly that can be likened to the U.S. Senate—aside from the UN's lack of power. There is also a Security Council, which, if stripped of its absolute power, could function like the U.S. House of Representatives to give the powerhouse nations some special representation. We must hope that such a dreadful catalyst never comes, but at least we'll have a general model to think about if it does.

Another possibility would be that some very immediate *external* threat could unite our dissident nations, but such a threat is difficult to imagine unless a predictably wayward comet might do the trick. In this improbable fantasy, all the nuclear nations would be impelled to expend their arsenals in order to save the planet. In another, the threat that might best unite our world of nations would lie purely in the realm of science fiction in the form of an imaginary alien political empire intent on interplanetary domination. Just as rivalrous male chimpanzees set aside their differences on patrol when attacking a stranger, it is quite predictable that our world of nations would unite under such putative conditions, and that it might do so if a real political threat of sufficient gravity were to arise.

Much closer to reality, climate changes leading to global starvation might unite us for a time. However, when such chips were *really* down, we might follow the path of a hunting band that is up against the wall and socially "atomize," with every nation looking out for itself and fearing its hungrier neighbors. Perhaps new epidemic diseases might galvanize cooperation, depending on the nature of the disease, and do so in a way that would induce enemies to become friends, so there's yet another "external threat" that might tend to unite us. But I believe that the most potent threat of all is that of nuclear annihilation, and so far, in my opinion, we've been willing to rely on a cross between an acephalous egalitarian hunter-gatherer political system and a despotic but essentially uncentralized chimpanzee system to contain this threat—without forming a really effective global moral community.

Perhaps our greatest hope should be placed in a world economic system that flourishes because of free trade, for as I've said this creates interdependencies that make serious conflicts costly in ways that are new. Another potentially positive factor, well worth mentioning, is global communication media. Eventually, global television and particularly the Internet may have some effect in significantly homogenizing our world culture and in thereby breaking down some of the cultural and religious diversities that work against trust among nations and can easily foster conflict. At the same time, certain shared aspects of world religions, including the Golden Rule, provide the potential for developing a greater moral community of interest. However, both modern communications and organized religion obviously can set us apart as well as help us to grow together, because either can tie in to ethnocentrism and foster xenophobia.

Still another major factor in at least *trying* to predict our future as a global community is simply the human political mind, which provides exactly the same potential that shaped earlier efficient moral communities in the form of egalitarian hunting bands—with an insistence on no centralized power, which made sense because to resolve serious conflicts avoidance was possible. In our later political evolution these same political and moral minds have created and accepted command and control as needed to make possible the centralized functions required to run much larger, sedentary societies as these develop. Again, our famous social and political flexibility has been at work, and up to the level of nations it has done its job well.

Flexibility means we are not by evolutionary design just lovers of equality. Indeed, in terms of human nature we seem to be just as capable of *following* leaders as we once were of getting rid of them generically. The basic political tendencies were present in Ancestral *Pan,* who had hierarchies with aggressively greedy alpha males that helped to keep down conflict with their forceful interventions—and therefore were at the same time resented and appreciated. Our very nature sets us up to be ambivalent about the exercise of power from above, and we're quite good both at holding down leaders and at appreciating command and control where we see this as being worthwhile or necessary to avert chaos, or where its image is one of generosity. This flexibility could be useful in the future, just as it was useful 45,000 years ago, and more, in often dangerous Late Pleistocene settings that because of large-game hunting generally favored a rampant version of egalitarianism at the expense of centralized governance.

Fearing undue domination, democratic nations resolve this ambivalence constitutionally, by keeping governmental powers checked and balanced. The global problem is that *trustworthy* checks and balances remain to be invented. Furthermore, the world seems to be simply too big—and perhaps too diversified—to ever agree upon a single leader unless the right and truly trusted "charismat" were to come along and gain the confidence of all. Even the hideously conflicted Balkans were united for decades by a charismatic Tito, while a highly respected George Washington managed to get a handful of quite disparate former British colonies off to the right start—even though as with the Balkans a civil war ensued. Unfortunately, it's far easier to visualize a *formal* world governmental structure, which at least in theory might work, than it is to create trust among competing and quite disparate nations with histories of rivalry and conflict.

I've said that the international system we actually live by looks like a cross between a band society—in which nobody wants to be bossed around—and a chimpanzee community, in which the big guys forcefully run the show and hog the resources but also serve usefully as effective and impartial peacemakers. I emphasize that globally, whoever fills the role of chief dominator is also expected to intervene *impartially* in conflicts and pacify them. For over half a century a major and perpetual problem faced by Superpower America has been its chosen role as a committed partisan supporter of an embattled but territorially aggrandizing Israel, which continues to make any really effective US role in mediation between Israelis and Palestinians extremely problematic. Unfortunately, this conflict provides the political engine that drives much of today's world conflict, as we begin to struggle through the second decade of the twenty-first century.

That's where we find ourselves, with our world of nations. Our ambivalent natures do provide us at least with an altruistic sense of empathy for others, which in the right contexts seems to be extendable to all of humanity. And we do have a sense of morality that helps us significantly in building very large national communities of interest which (as with our American political union) may with difficulty endure—but which (as with the Soviet Union) also may fall apart. Morally, we do behave in many contexts as a world community that judges individual nations in terms of right and wrong; we even have functioning international courts, though their jurisdiction is not accepted universally. These are significant signs of some limited unification for the common good—if not some kind of a serious and binding political confederation.

At the same time, we're faced with hundreds of sizable, often well-armed national sovereignties, which can become heavily involved with ethnocentrism—and sometimes with raw xenophobia—to create global situations of sharply competing alliances and mistrust. We're also susceptible to morally based ideologies, some of them destructive to world cooperation and some, like the ongoing nuclear taboo, highly benign. Fortunately, ideologies that promote generosity are grounded deeply in our nature, and in our cultures as well. The Golden Rule has been a human universal for at least 45,000 years, and this continues today. Such ideologies may help significantly in facilitating a world order that could be made to be more trusting and interdependent—and hence at least somewhat less competitive and dangerous.

In its specifics our future history is unpredictable at the global level, even though supposedly the past can predict the future. What the *recent* historical past tells us, in statistical terms, is that death and destruction owing to warfare may well be declining. But does that mean that ultimate risks are declining as well? Our much deeper *evolutionary* past provides a different means of prediction, and what it seems to be telling us, here, is that human nature gives us a great deal to build upon, but also a great deal to fear. In solving future problems, I believe that it's important to know about the basics of what we have to work with. I also believe that as time takes us forward, the great similarities between the world community of nations and an LPA band can provide us with some important food for thought, as do the several signal differences I've just discussed.

In searching for our moral origins in the preceding chapters, we've taken an enormous journey. And perhaps what we've learned about the Late Pleistocene can teach us a bit more about the global problems we'll be facing as this journey continues. Our moral capacity is part of the potential we carry into this future, and one thing we'll have to work with, whenever we're ready to move in the direction of creating a less risky global moral community, is the same moral nature that archaic and then culturally modern humans evolved for us in the Late Pleistocene. Combine this with our rather remarkable political inventiveness, and humanity may have some major reason for hope.

Acknowledgments

This book consolidates a variety of research interests pursued over the past three decades. The groundwork for this analysis was done under fellowships from the John Simon Guggenheim Foundation, the School of Advanced Research in Santa Fe, and the National Endowment for Humanities, and under grants from the John Templeton Foundation, the Harry Frank Guggenheim Foundation, and the L. S. B. Leakey Foundation.

I am grateful to the Gombe Stream Research Centre for support of relevant fieldwork on primates in Tanzania, while for help in past work that figured importantly in the writing of this book I thank Jane Ayers, Nigel Barradale, Donald Black; Deborah Boehm, Michael Boehm, Sam Bowles, Sarah Brosnan, Rose Ann Caiola, James Francis Doyle, Carol Ember, Dean Falk, Jay Feierman, Jessica Flack, Roger Fouts, Doug Fry, Herb Gintis, Michael Gurven, Jonathan Haidt, Kristin Howard, Hilly Kaplan, Raymond C. Kelly, Bruce Knauft, David Krakauer, Don Lamm, Frank Marlowe, Michael Mcguire, Steven Morrissey, Martin Muller, Lluis Oviedo, John Price, Karl Recktenwald, Pete Richerson, Alice Schlegel, Jeffrey Schloss, Doron Schultziner, Craig Stanford, Mary Stiner, Jonathan Turner, Frans de Waal, Nicholas Wade, Paul Wason, Mary Jane West-Eberhard; Andy

Whiten, Polly Wiessner, David S. Wilson, Michael L. Wilson, and Richard Wrangham.

More recently I must thank Sam Bowles, Jean Briggs, Jessica Flack, Herb Gintis, Jonathan Haidt, Kim Hill, Jim Hopgood, Mel Konner, Deirdre Mullane, Randolph Nesse, Judy Vinegar, Polly Wiessner, and Frans de Waal for detailed comments on the present manuscript, and I also wish to thank Richard Wrangham for sharing an unpublished book manuscript in which the importance of prehistoric capital punishment's acting on gene pools was emphasized, as an instance of "autodomestication."

At Basic Books I wish to thank their excellent editorial staff for the help they have offered, including T. J. Kelleher, Tisse Takagi, and Collin Tracy.

Kristin Howard's work as hunter-gatherer research assistant has been impeccable, and I thank my daughter, Jennifer Morrissey, for her work in creating the index. I must give special thanks to Jane Goodall, who trained me in ethological field techniques. In addition I thank my agent, Deirdre Mullane, for the substantial help and support that helped this writing project to succeed, and my colleague Don Lamm for encouraging the project over a period of several years.

Finally I must pay homage to two late mentors. One was anthropologist Paul J. Bohannan, who encouraged me to take up primatology as a cultural anthropologist; the other was psychologist Donald T. Campbell, to whom the book is dedicated. It was Don Campbell who suggested that I leave linguistic anthropology and become an evolutionist.

Notes

NOTES TO CHAPTER 1: DARWIN'S INNER VOICE

1. See Richards 1989.
2. See Darwin 1859.
3. See ibid.
4. See ibid.
5. See ibid.
6. See Campbell 1975.
7. See Campbell 1965.
8. See Malthus 1985 (1798).
9. See Spencer 1851.
10. See Lyell 1833.
11. See Flack and de Waal 2000.
12. See Darwin 1982 (1871), 71–72.
13. See Campbell 1975; see also Alexander 1974 and Wilson 1975.
14. See Hamilton 1964.
15. See Darwin 1982 (1871).
16. See Williams 1966.
17. See Wilson 1975.
18. See West et al. 2007.
19. See Wilson 1975 for an assessment of how genetic preparations make certain behaviors very easy to learn and how genetic leashes constrain behavior at the level of phenotype.
20. See ibid.

21. See Boehm and Flack 2010.

22. See Boehm 1979 and Boehm 2009; see also Sober and Wilson 1998.

23. See Boehm 2004a and Boehm 2008a.

24. See Kelly 1995.

25. See Alexander 1987.

26. See Boehm 2008b.

27. See Wilson and Wilson 2007.

28. See Darwin 1982 (1871), 98.

29. See Wilson and Wilson 2007.

30. See Darwin 1865 and Darwin 1982 (1871).

31. See Alexander 1987.

32. See Boehm 1997.

33. See West-Eberhard 1983; see also Nesse 2007, Bowles and Gintis 2011, Boehm 1978, Boehm 1991a, and Boehm 2008b.

34. See Boehm 1991b.

35. See Campbell 1965.

36. See Mayr 1988.

37. In this sense, humans have actually constructed part of the environment they are adapted to. See, for example, Boehm and Flack 2010; see also Laland et al. 2000.

38. Darwin talked about "sympathy" as a basis for humans and other social animals sacrificing their own interests to assist others, be they kin, nonkin, or even members of other species. Today, a number of researchers have investigated empathy as a way of understanding the emotions and cognitions that impel us to feel for others and assist them. There appear to be various types of empathy as well as differing academic definitions (see, for example, Batson 2011, Flack and de Waal 2000, Preston and de Waal 2002, and de Waal 2009), so basically I shall stay with Darwin's more general-purpose term: sympathy.

NOTES TO CHAPTER 2: LIVING THE VIRTUOUS LIFE

1. See Piers and Singer 1971 for an interesting treatment of anthropological research on shame and guilt.

2. See Casimir and Schnegg 2002.

3. See Darwin 1982 (1871).

4. There is a fascinating study, conducted at a Siberian fox farm, that supports this view. See Trut et al. 2009.

5. See Lindsay 2000.

6. See Damasio 2002.

7. See Damasio et al. 1994. The Phineas Gage case history is particularly evocative for me because my maternal grandfather, William Askey of the Erskine clan, whose father emigrated from Scotland, was a foreman with the Baltimore and Ohio Railroad in western Maryland. He wore a glass eye that replaced the one he had lost in a similar but less traumatic accident.

8. See Damasio 2002.

9. See Hare 1993.

10. See Parsons and Shils 1952; see also Gintis 2003.

11. See Darwin 1982 (1871).

12. See Kiehl 2008; see also Kiehl et al. 2006.

13. See Freud 1918.

14. See Frank 1988.

15. See Alexander 1987, 102.

16. See Faulkner 1954.

17. See Batson 2009.

18. See Dunbar 1996.

19. See Boehm 1993.

20. See Turnbull 1961.

21. See Boehm 2004b.

22. See Turnbull 1961. The extensive quotations are to be found in Colin Turnbull's (1961) *The Forest People*, Chapter 5, "The Crime of Cephu, the Bad Hunter," 94–108.

23. See Lee 1979, 244.

24. See ibid., 246.

25. See Durham 1991.

26. See Wiessner 2002.

NOTES TO CHAPTER 3: OF ALTRUISM AND FREE RIDERS

1. See Campbell 1975; see also Boehm 2009.

2. See Campbell 1972. See Sober and Wilson 1998, 142–149, for a fuller treatment of how altruism can be socially "amplified," and Boehm 2004a.

3. See Fehr and Gächter 2002; see also Henrich et al. 2005 and Hammerstein and Hagen 2004. Generous moves in experimental games can lead to reciprocation in kind.

4. See Boehm 2008b.

5. This holds for the !Kung according to Polly Wiessner (personal communication).

6. See Gurven et al. 2001.

7. Ibid; see also Alexander 1987.

8. See Smith and Boyd 1990 and Wiessner 1982.

9. See Malinowski 1922.

10. See Gurven et al. 2001.

11. See Service 1975.

12. See Campbell 1975.

13. See Sober and Wilson 1998.

14. See Campbell 1972, 1975.

15. See, for instance, Aberle et al. 1950 for an academic view of what it takes to keep a society going.

16. See Eibl-Eibesfeldt 1982; see also Gintis 2003 and Simon 1990.

17. See Boehm 2009.

18. See Darwin 1982 (1871).

19. See Williams 1966.

20. See Sherman et al. 1991.

21. See Wilson 1975.

22. See Irons 1991 for one answer to this question; see also Lewontin 1970.

23. See Alexander 1987; see also Black 2011, Boehm 2000, Boehm 2008b, Boyd and Richerson 1992, and Simon 1990.

24. See, for example, Preston and de Waal 2002; see also Flack and de Waal 2000.

25. See Batson 2009 and de Waal 2009 for technical treatments of human sympathy that use "empathy" as a more carefully defined, scientific term that has found its way into our everyday vocabularies.

26. See Darwin 1982 (1871).

27. I borrow this terminology from Donald T. Campbell's writings on altruism. See Campbell 1975.

28. See Williams 1966, 205.

29. See ibid., 203–204.

30. See ibid. For an example of this theory being put to use, see Boehm 1981. This application involves macaque monkeys, and the hypothesis is that costly alpha-male pacifying interventions in adult fights are altruistic and that they are an extension of females' stopping fights among offspring, which pay off handsomely because of kin selection. The theory is that the alpha's

interventions—those that protect unrelated adults—are genetically piggybacking on the maternal interventions that protect offspring.

31. There are a number of ways of looking at altruism (see West et al. 2007), and here I have organized this effort in terms that should be both unambiguous and understandable to general readers. In doing so, I have restricted the meaning of altruism to costly generosity that is *extrafamilial.*

32. See Hill et al. 2011.

33. See, for example, Gintis 2003 and Simon 1990; see also Alexander 2006.

34. In biology, pleiotropy means that the same gene can have two or more disparate effects.

35. Piggybacking means that as an instance of pleiotropy, a useful trait may "carry" a somewhat deleterious trait that is set up by the same gene. This can work as long as the useful trait is strongly selected and the piggybacking trait isn't too costly. See Gintis 2003.

36. See Bowles 2006; see also Sober and Wilson 1998.

37. See Mayr 2001.

38. See Wilson 1975; see also ibid.

39. See Alexander 1979; see also Wilson 1975 and Trivers 1971.

40. See Bowles 2006 and Bowles 2009.

41. See Trivers 1971.

42. See, for instance, Allen-Arave et al. 2008.

43. See Stevens et al. 2005.

44. See Alexander 1987.

45. See Kaplan and Hill 1985; see also Allen-Arave et al. 2008 for an example having to do with nepotistic food transfers, and Kaplan and Gurven 2005 for an overview.

46. See, for example, Alexander 2006, Brown et al. 2003, and Hammerstein and Hoekstra 2002.

47. See Marlowe 2010.

48. See Alexander 1987.

49. See ibid.

50. See, for instance, Bird et al. 2001; see also Zahavi 1995. Costly signaling is more narrowly conceived than Alexander's original idea of selection by reputation, which is gossip based and involves far more than hunting prowess and can involve signals that are not indicative of fitness.

51. See Gurven et al. 2000; see also Henrich et al. 2005, Kaplan and Hill 1985, Nowak and Sigmund 2005, Wiessner 1982, and Wiessner 2002.

52. See Alexander 1987.

53. See Boehm 1997 and Boehm 2000.

54. See Williams 1966; see also Trivers 1971.

55. See Wilson and Wilson 2007.

56. See Alexander 1987.

57. See Cosmides et al. 2005; see also Trivers 1971, Williams 1966, and Wilson 1975.

58. See Boehm 1997 and Cummins 1999.

59. See Ellis 1995; see also Betzig 1986.

60. See Boehm 1999 and Boehm 2004a.

61. See Boehm 1999; see also Erdal and Whiten 1994.

62. See Cosmides et al. 2005.

63. See Boehm 1993; see also Erdal and Whiten 1994.

64. See Boehm 1997.

65. See Frank 1995.

66. See Fehr et al. 2008.

67. See Boehm 1999.

68. See Williams 1966.

69. See Boyd and Richerson 1992.

70. See Boehm 1993.

71. See Wiessner 2005a and 2005b; see also Lee 1979.

72. See Briggs 1970, 44.

73. See ibid., 47.

74. See ibid.

75. See ibid.

76. See Boehm 1993.

77. See also Bowles 2006.

78. See Hill et al. 2011.

79. See Bowles 2006 and Bowles 2009; see also Choi and Bowles 2007.

80. In the past, I have tried to give group selection every benefit of the doubt at a time when such theory was far more controversial than it is today; see, for example, Boehm 1993, Boehm 1996, Boehm 1997, and Boehm 1999. The free-rider suppression that is emphasized in this book addresses a major problem with group selection models as these were originally attacked by Williams (1966) and should add to the viability of group selection theory as applied to humans.

81. See West-Eberhard 1979, West-Eberhard 1983, and Wolf et al. 1999 for a broader perspective.

82. I thank Randolph Nesse for the suggestion that no single mechanism is likely to fully explain human altruism; see also Gurven and Hill 2010.

NOTES TO CHAPTER 4: KNOWING OUR IMMEDIATE PREDECESSORS

1. See Klein 1999.
2. See Kelly 1995; see also Service 1975.
3. See Burroughs 2005.
4. See Gould 1982.
5. See Dyson-Hudson and Smith 1978.
6. See Kelly 2000.
7. See Bowles 2006.
8. Unfortunately, the most detailed study, by Turnbull 1972, is of relocated hunter-gatherers for whom no baseline study exists.
9. See Balikci 1970; see also Mirsky 1937 and Riches 1974.
10. See Lee 1979.
11. See, for instance, Kelly 1995; see also Service 1975. Lawrence Keeley (1988) is an exception; he compiled a list of ninety-four economically independent foragers to investigate variations in prehistoric behavior with respect to population pressure and socioeconomic complexity.
12. See Steward 1955.
13. See Boehm 2002 and Boehm 2012.
14. See Binford 2001.
15. See also Keeley 1988.
16. See Boehm 2012.
17. See Klein 1999.
18. See ibid.
19. See Marlowe 2005 for some of the information just cited.
20. See Klein 1999.
21. See Hill et al. 2011.
22. See Kelly 1995.
23. See Potts 1996.
24. See Gould 1982; see also Steward 1938.
25. See McBrearty and Brooks 2000.
26. See Fleagle and Gilbert 2008.
27. See Boehm 1999.
28. See Lee 1979.

29. See Boehm 2008b; see also West-Eberhard 1979 and West-Eberhard 1983.

30. See, for instance, Wilson 1978.

NOTES TO CHAPTER 5: RESURRECTING SOME VENERABLE ANCESTORS

1. See Flack and de Waal 2000.

2. See Watson and Crick 1953.

3. See Ruvolo et al. 1991.

4. See Boehm 2004b.

5. See Wrangham 1987.

6. See Brosnan 2006.

7. See Wrangham 1987.

8. Homology means, for instance, that two species exhibit a similar behavior because their genetic makeups are similar, and in turn this means that the behaviors are based on similar psychological mechanisms.

9. See Wrangham and Peterson 1996.

10. I first heard this term being used by my colleague Martin Muller, who is a primatologist at the University of New Mexico.

11. See Barnett 1958 and Lore et al. 1984.

12. See Wrangham and Peterson 1996.

13. See Boehm 1999.

14. See Boehm 1993.

15. See Erdal and Whiten 1994 for the origin of the term "counter-domination."

16. See Hrdy 2009.

17. See Klein 1999.

18. See Keenan et al. 2003; see also Kagan and Lamb 1987.

19. See Malinowski 1929.

20. See Boehm 2000.

21. See Boehm 1999.

22. Cultural transmission is discussed by Boyd and Richerson 1985, while gene-culture evolution is explored by Durham 1991, with in-depth ethnographic exemplification.

23. See Darwin 1982 (1871).

24. See Damasio 2002.

25. See Gallup et al. 2002.

26. See Bearzi and Stanford 2008.

27. See Mead 1934.

28. See Gallup et al. 2002.

29. See, for instance, Bearzi and Stanford 2008.

30. See Gardner and Gardner 1994.

31. See Menzel 1974.

32. See de Waal 1982.

33. See Goodall 1986.

34. See Whiten and Byrne 1988.

35. See Diamond 1992.

36. See Durkheim 1933.

37. See de Waal and Lanting 1997.

38. See Lee 1979.

39. See Kano 1992.

40. See Parker 2007.

41. See de Waal 1982.

42. See Goodall 1986.

43. See Goodall 1992 and Nishida 1996; see also de Waal 1986 and Ladd and Mahoney 2011.

44. See de Waal 1996 and Flack and de Waal 2000; see also McCullough et al. 2008.

45. See de Waal 1996, 91–92.

46. See, for example, Gardner and Gardner 1994; see also Savage-Rumbaugh and Lewin 1994.

47. See Goodall 1986.

48. With respect to apes' learning American Sign Language, see Gardner and Gardner 1994 and Patterson and Linden 1981.

49. See Savage-Rumbaugh and Lewin 1994.

50. See Temerlin 1975, 120–121.

51. See Boehm 1980.

52. See de Waal 1982.

53. See de Waal 1996.

54. See Fouts 1997, 156.

55. See also Whiten and Byrne 1988.

56. See Savage-Rumbaugh et al. 1998, 52.

57. See ibid., 52–53.

58. See Fouts 1997, 151–152.

59. See Patterson and Linden 1981.

60. See ibid., 39.

61. See Flack and de Waal 2000; see also Preston and de Waal 2002 for a discussion of different types of empathy.

NOTES TO CHAPTER 6: A NATURAL GARDEN OF EDEN

1. See Cavalli-Sforza and Edwards 1967.
2. See Wrangham and Peterson 1996.
3. See Pinker 2011.
4. See Bowles and Gintis 2011.
5. See, for instance, Goodall 1986.
6. See, for example, Burch 2005.
7. See Kano 1992.
8. See ibid.
9. See Furíuchi 2011, see also de Waal and Lanting 1997.
10. See Bowles 2006 and Bowles 2009.
11. See Sober and Wilson 1998.
12. See LeVine and Campbell 1972.
13. See ibid.
14. See Keeley 1996.
15. See Bowles 2006.
16. See ibid.
17. See Noss and Hewlett 2001; see also Mirsky 1937.
18. See Burroughs 2005.
19. See Stanford 1999.
20. See Boesch and Boesch-Achermann 1991.
21. See Blurton Jones 1991.
22. See Boehm and Flack 2010.
23. See Wrangham 1999.
24. See Watts and Mitani 2002.
25. See Boesch and Boesch-Achermann 1991 and Boesch and Boesch-Achermann 2000.
26. See Byrne 1993 for an assessment of empathy in other primates.
27. See Stanford 1999.
28. See Hohmann and Fruth 1993.
29. See Hrdy 2009.
30. See Kelly 1995.
31. See Peterson 1993.
32. See ibid.
33. See Winterhalder 2001.
34. See Winterhalder and Smith 1981; see also Winterhalder 2001.
35. See Winterhalder and Smith 1981.

36. See Kaplan and Hill 1985.
37. See Boehm 1982.
38. See Boehm and Flack 2010.
39. See Beyene 2010.
40. See Klein 1999.
41. See ibid.
42. See ibid.
43. See ibid.
44. See Thieme 1997.
45. See Stiner 2002.
46. See Boehm 1999.
47. See Ellis 1995.
48. See Nishida 1996.
49. See Goodall 1992.
50. See Parker 2007.
51. See Kano 1992.
52. The philosopher Karl Popper judges theories by their "falsifiability," which means that they must be couched in terms that are susceptible of testing. In doing so, he makes some special allowances for the uniqueness of Darwinian explanations, which in their larger aspects are testable mainly in terms of their general plausibility in competition with other explanations. See Popper 1978.
53. See Campbell 1975.
54. See Whallon 1989.
55. See Wrangham and Peterson 1996.
56. See Stiner 2002.
57. See Hawks et al. 2000.
58. See Bunn and Ezzo 1993; see also Speth 1989.
59. See Boehm 2004b; see also Hawkes 2001.
60. See Whallon 1989; see also Knauft 1991.
61. See Pericot 1961.
62. See Kelly 2000.
63. See Lee 1979; see also Kelly 2005.
64. See Knauft 1991.
65. See Stiner et al. 2009.
66. See Klein 1999.
67. See Boehm 2004b; see also Boehm 1982 and Boehm 2000.
68. See Kelly 1995.
69. See Eldredge 1971.

70. See Boehm and Flack 2010.

71. For a broad evolutionary view of problem solving, see Dewey 1934. John Dewey's views on evolution have yet to receive the attention they deserve.

72. See Wilson 1978.

73. See West-Eberhard 1983.

74. See Darwin 1982 (1871).

75. See Fisher 1930; see also Nesse 2000 and Nesse 2007.

76. See Campbell 1965; see also Campbell 1975.

77. See Trivers 1971. In this seminal article sociobiologist Robert Trivers led the way, not only in considering moralistic aggression as a force that acted on human gene pools, but also in analyzing pairwise cooperation from both a genetic and a psychological perspective.

78. See Alexander 1979 and Alexander 1987.

79. See West-Eberhard 1979 and West-Eberhard 1983.

80. See, for instance, Wiessner 1996 for her most recent treatment of !Kung safety nets.

81. See Alexander 1987, 94.

82. See Otterbein 1988.

83. See Wrangham 2001; see also Wrangham and Peterson 1996.

84. See Boehm 1999, 253–254.

85. See Voland and Voland 1995.

86. My Unitarian mother wanted me to make my own choices about religion, so she sent me to a variety of Sunday schools.

87. See Haile 1978.

88. See Haidt 2007.

89. See Haidt 2003.

90. See Greene 2003.

91. See Alexander 1987.

NOTES TO CHAPTER 7: THE POSITIVE SIDE OF SOCIAL SELECTION

1. See Alexander 1979 and Alexander 1987.

2. See Trivers 1972.

3. See Campbell 1975; see also Neusner and Chilton 2009.

4. See Sullivan 1989.

5. See Campbell 1975.

6. See Alexander 1987.

7. See Boehm 1986.
8. See Durkheim 1933.
9. See Gurven et al. 2000.
10. See de Waal 2009; see also Flack and de Waal 2000 and Hrdy 2009.
11. See Kelly 1995.
12. See Balikci 1970.
13. See Wilson 1999.
14. See Zahavi 1995.
15. See ibid.
16. See Darwin 1982 (1871); see also Zahavi 1995.
17. See, for instance, Lee 1979.
18. See Keeley 1988; see also Kelly 1995.
19. Frank Marlowe, at Cambridge University in England, has already published on this basis. See Marlowe 2005. Kim Hill, at the University of Arizona at Tempe, is also working with a sizable database. See Hill et al. 2011. These databases focus on subsistence techniques and main features of social organization.
20. See Campbell 1972 and Campbell 1975; see also Brown 1991.
21. See Campbell 1975.
22. See Sober and Wilson 1998.
23. See Boehm 2008b.
24. See Alexander 1987.
25. See ibid.
26. See Marlowe 2005.
27. See Balikci 1970.
28. The analogy comes to mind because I once observed a Hopi Snake Dance, where the small desert rattlers that dancers carried in their mouths had secretly had their fangs removed. I didn't know about it at the time, and this lack of knowledge contributed to an extreme case of culture shock on my part because a young dancer was "bitten" repeatedly on the cheek and the other dancers were simply taking this for granted.
29. See Wrangham and Peterson 1996.
30. See Johnson and Krüger 2004; see also Wade 2009.
31. See Johnson and Krüger 2004.
32. See Dawkins 1976, Ridley 1996, and Wright 1994.
33. See Ghiselin 1974.
34. See Boehm 1999.
35. See Ghiselin 1974, 247.

36. See Panchanathan and Boyd 2004.

37. See, for example, Fehr and Gächter 2002.

38. See Henrich et al. 2005.

39. See Boyd et al. 2003; see also Fehr 2004, Fehr and Gächter 2002, Kollock 1998, Panchanathan and Boyd 2004, and Price et al. 2002.

40. See Guala in press; see also Boehm in press.

41. See Lee 1979.

42. See Boehm 2011.

43. See West-Eberhard 1979.

44. See Boehm 1982; see also Alexander 1987.

45. See Zahavi 1995; see also Bird et al. 2001.

46. See Hrdy 2009.

47. See Boehm 2004a.

NOTES TO CHAPTER 8: LEARNING MORALS ACROSS THE GENERATIONS

1. See Turnbull 1961.

2. See Durkheim 1933.

3. See, for instance, Elkin 1994.

4. See Coser 1956.

5. See Boehm 1999.

6. See Boehm 1983 and Boehm 1986.

7. For the Bushmen, see, for instance, Lee 1979, Heinz 1994, and Silberbauer 1981. For the Inuit, see, for instance, Balikci 1970 and Briggs 1970.

8. See Thomas 1989, Lee 1979, Wiessner 1982, Wiessner 2002, and Draper 1978.

9. See Shostak 1981.

10. See Rasmussen 1931 and Balikci 1970.

11. See Briggs 1970.

12. See Briggs 1998.

13. See Parsons and Shils 1952.

14. See Simon 1990.

15. See Gintis 2003.

16. See Waddington 1960 and Campbell 1975.

17. See, for instance, Eisenberg 2006 and Turiel 2005; see also Konner 2010.

18. See Kagan 1981; see also Kagan and Lamb 1987.

19. See Campbell 1975.

20. Robert Kelly published this example in his comprehensive 1995 book on hunter-gatherers. See also Leacock 1969.

21. See Leacock 1969, 13–14.

22. See Stephenson 2000.

23. See Westermarck 1906.

24. See ibid., 118.

25. See Draper 1978, 42; italics added.

26. See Whiting and Whiting 1975.

27. See Boehm 1972; see also Boehm 1980.

28. See Gallup et al. 2002.

29. See ibid., and Kagan and Lamb 1987.

30. See, for example, Warneken et al. 2007, which compares spontaneous altruism in human children and in young chimpanzees.

31. See Greene 2007. To explain further, the experiments involve a subject's having to choose between passively letting five persons die in a runaway trolley and actively killing a sixth person to save the other five. In the first hypothetical scenario, you merely throw a switch so that the trolley will run into a bystander on the track, which saves the five passengers' lives; in the second you involve yourself much more actively by pushing the fat guy off a bridge so that he falls in the path of the trolley and stops it.

32. See Briggs 1994.

33. See Briggs 1982, 118–119.

34. See ibid., 120–121.

35. See ibid., 121.

36. See Boehm 1989 and Boehm 1999.

37. See Briggs 1998.

38. See Hewlett and Lamb 2006.

39. See Konner 2010.

40. See ibid.

41. The quotations on pages 228–233 are taken from Shostak 1981, 46–57.

42. See ibid., 56.

43. See Durham 1991.

44. See, for instance, Freud 1918.

45. See, for instance, Wilson 1975.

46. See Konner 2010.

NOTES TO CHAPTER 9: WORK OF THE MORAL MAJORITY

1. See Haviland 1977.

2. See Wiessner 2005a and Wiessner 2005b.

3. The Aleuts executed serious gossips whose words were harming their communities. See Jones 1969.

4. See Boehm 1986.

5. See Bogardus 1933 and Boehm 1985.

6. See Lee 1979.

7. See Briggs 1970.

8. See ibid.

9. See Boehm 1999, 57–58.

10. This extreme form of ostracism was mentioned in Table IV, page 198.

11. See Boehm 1985.

12. See Durham 1991.

13. See Haidt 2007.

14. See Westermarck 1906.

15. See Wolf and Durham 2004.

16. See Goodall 1986.

17. See Cantrell 1994.

18. See Balikci 1970, 191.

19. See Thomas 1989.

20. See Knauft 1991.

21. See Lee 1979.

22. See ibid., 372–373.

23. See Boehm 2011 and Knauft 1991.

24. See Fry 2000; see also von Furer-Haimendorf 1967 and Knauft 1991.

25. See Lee 1979.

26. See Boehm 2004b. Subsequently, Polly Wiessner explained to me exactly how this works with the Bushmen.

27. See Lee 1979.

28. See Draper 1978, 46.

29. See van den Steenhoven 1957, van den Steenhoven 1959, and van den Steenhoven 1962.

30. See Balikci 1970, 195–196.

31. See ibid.

32. See Lee 1979.

33. See Balikci 1970.

34. See Knauft 1991.

35. See Boehm 2007 and Boehm 2011.

36. See Lee 1979.

37. See ibid., 394–395.

NOTES TO CHAPTER 10: PLEISTOCENE UPS, DOWNS, AND CRASHES

1. See Balikci 1970.

2. Actually, there are fourteen separate body parts that can be exchanged in this way, but only seven of them involve long-term partnerships. See ibid.

3. See ibid.

4. See, for instance, Binford 1978.

5. See Balikci 1970.

6. See Lee 1979.

7. See Briggs 1970.

8. See Peterson 1993.

9. Again, see Sober and Wilson 1998 for a discussion of this important phenomenon. See also Boehm 2004a.

10. See Peterson 1993.

11. See Keely 1988.

12. See Gould 1982.

13. See, for instance, Balikci 1970.

14. See Laughlin and Brady 1978.

15. See Testart 1982.

16. See, for instance, Balikci 1970.

17. Leibig's Law of the Minimum is often cited without further attribution. Actually, Baron Justis von Leibig merely popularized the idea of Karl Phillip Sprengel, a German agronomist who came up with the idea that the most scarce element in an environment would limit the success of a species. For instance, in a serious drought, this would be water. See Sprengel 1839.

18. See Hawks et al. 2000.

19. See Bowles 2006.

20. See Burroughs 2005.

21. See Balikci 1970.

22. See ibid.

23. See ibid.

24. See Gould 1982.

25. See Keeley 1988.
26. See Potts 1996.
27. See Boehm 1996.
28. See Balikci 1970.
29. See Peterson 1993.
30. See Shostak 1981, 44.
31. See ibid.
32. Quotes from *Nisa* on pages 282–288 taken from Shostak 46–54.
33. See Shostak 1981, 323.
34. See Wiessner 1982.
35. See Lorenz 1966.

NOTES TO CHAPTER 11: TESTING THE SELECTION-BY-REPUTATION HYPOTHESIS

1. See also de Waal 2008 and de Waal 2009.
2. See Gurven et al. 2000, 266; see also Gurven 2004.
3. See Gurven et al. 2000.
4. See also Woodburn 1982.
5. See Bird et al. 2001; see also Hawkes 1991 and Smith 2004.
6. Hawkes 1991, for example, emphasizes this in evaluating consequences of hunting success.
7. See Kelly 1995, 164–165; see also Bird-David 1992 and Myers 1988.
8. See Marlowe 2004.
9. See Woodburn 1979.
10. For the Hadza, see ibid.
11. See Shostak 1981, 116.
12. See Sober and Wilson 1998.
13. See Hill et al. 2011. Actually, a total of thirty-two foraging societies were sampled, of which about a third fit the technical criteria used here for being "Late Pleistocene appropriate or LPA."
14. See ibid.
15. An exception is that as currently defined, kin selection and group selection appear to have overlapping applications because costly acts of generosity within groups of kin can be explained by either model. Alexander's consideration of group-level selection taking place within kin units reflected this ambiguity. See, for instance, Wilson and Sober 1994 and Sober and Wilson 1998. A type of group selection that also could be affecting the social behaviors that are

influencing genetic outcomes is cultural group selection. See Richerson and Boyd 1999.

16. See Boehm 1993 for a study that identifies qualifications for being chosen as a leader in egalitarian societies.

17. See Alexander 1987; see also Marlowe 2010, Figure 7.4.

18. See Wiessner 1982 and Wiessner 2002.

19. See Wilson and Dugatkin 1997.

20. Assortative mating is a major field of study for humans and other species, the idea being that individuals may choose as mates others who are similar physically or behaviorally. In this context, Wilson and Dugatkin 1997, have explored the possibility that altruists might be choosing other altruists (see also Hamilton 1975); their interest is in looking at group selection with respect to whether assortative choices could be enhancing the strength of group effects by increasing between-group variation, whereas my interest here is in whether such effects could be empowering selection by reputation that takes place *within* a group.

21. It doesn't matter if the signal is costly or not. For instance, if two hard-working individuals choose each other, the signals are noncostly, but they are good indicators of superior genetic quality. If two altruists join up, the signals do have costs, but over time these costs will be mutually compensated because their cooperation will outclass the cooperation of a stingy pair.

22. See Fisher 1930. See also Nesse 2007 and Nesse 2010, who has discussed runaway selection in the specific case of altruism in humans and has emphasized its likely importance.

23. See Alexander 2005, 337–338.

24. See Mirsky 1937.

25. See Williams 1966.

26. See Tiger 1979.

NOTES TO CHAPTER 12: THE EVOLUTION OF MORALS

1. See Mayr 1983 and Mayr 2001.

2. See Boehm and Flack 2010 and Laland et al. 2010.

3. See Midgley 1994, 118–119.

4. See Klein 1999. It is worth noting that power scavenging could have produced enormous whole carcasses with a surfeit of meat, but also "leftovers" with only enough meat for higher-ranking individuals if they were aggressively eating their fill.

5. See Campbell 1972 and Campbell 1975.

6. See Huxley 1894 and Spencer 1851.

7. See Nietzsche 1887.

8. See Breasted 1933.

9. The original date of publication for this precocious work was 1906.

10. See ibid.

11. See Wilson 1975.

12. See Wilson 1978.

13. See Ridley 1996.

14. See Wright 1994.

15. See Wilson 1993.

16. See Shermer 2004.

17. See Mayr 1988 and Mayr 1997; see also Wilson and Sober 1994 and Sober and Wilson 1998.

18. See Hauser 2006.

19. See Katz 2000.

20. See Flack and de Waal 2000; see also Preston and de Waal 2002.

21. See Boehm 2000.

22. See Sober and Wilson 1998.

23. See Sober and Wilson 2000.

24. See Skyrms 2000.

25. But see Dubreuil 2010.

26. See Krebs 2000.

27. See Rapoport and Chammah 1965.

28. See Frank 1988.

29. See Fehr and Gächter 2004; see also Fehr et al. 2008.

30. See Bowles and Gintis 2004.

31. See Bowles 2006 and Bowles and Gintis 2011.

32. See Bowles 2009; see also Keely 1996 and Kelly 2000.

33. See, for instance, Mithen 1990.

34. I thank Frans de Waal for a discussion of this matter.

35. See, for instance, Nowak et al. 2010 and Wilson and Wilson 2007.

36. See Campbell 1975.

37. On Hutterite communities, see Sober and Wilson 1998.

38. See Hrdy 2009.

39. Chimpanzees gang up against leopards. See Boesch 1991 and Byrne and Byrne 1988. I also have a videotape of a sixteen-foot python being harassed by the Gombe chimpanzees. It seems likely that bonobos have a similar

mobbing capacity, but with far less field study, this has not been witnessed so far.

40. See Alexander 1987.

41. See ibid.

42. See Sober and Wilson 1998; see also Boehm 2008b and Boehm 2009.

43. See Boehm 1976, Boehm 1991a, and Boehm 2008b.

44. See Wilson 1975.

45. See, for example, Harpending and Rogers 2000 and Hawks et al. 2000.

46. See Burroughs 2005.

47. See Popper 1978.

NOTES TO EPILOGUE: HUMANITY'S MORAL FUTURE

1. See Boehm 2003.
2. See Pinker 2011.
3. See Boehm 2003.
4. See Pinker 2011.
5. See Wrangham 1999.
6. See Woodburn 1982.

REFERENCES

Aberle, D. F., Cohen, A. K., Davis, A. K., Levy, M. J., and Sutton Jr., F. X. 1950. The functional prerequisites of a society. *Ethics* 60:100–111.

Alexander, R. D. 1974. The evolution of social behavior. *Annual Review of Ecology and Systematics* 5:325–384.

——. 1979. *Darwinism and human affairs*. Seattle: University of Washington Press.

——. 1987. *The biology of moral systems*. New York: Aldine de Gruyter.

——. 2005. Evolutionary selection and the nature of humanity. In *Darwinism and philosophy*, eds. V. Hosle and C. Illies. Notre Dame, IN: University of Notre Dame Press.

——. 2006. The challenge of human social behavior: Review of Hammerstein, genetic and cultural evolution of cooperation. *Evolutionary Psychology* 4:1–32.

Allen-Arave, W., Gurven, M., and Hill, K. 2008. Reciprocal altruism, rather than kin selection, maintains nepotistic food transfers on an Aché reservation. *Evolution and Human Behavior* 29:305–318.

Balikci, A. 1970. *The Netsilik Eskimo*. Prospect Heights, IL: Waveland Press.

Barnett, S. A. 1958. An analysis of social behavior in wild rats. *Proceedings of the Zoological Society of London* 130:107–151.

Batson, C. D. 2009. These things called empathy: Eight related but distinct phenomena. In *The social neuroscience of empathy*, eds. J. Decety and W. Ickes. Cambridge, MA: MIT Press.

——. 2011. *Altruism in humans*. New York: Oxford University Press.

Bearzi, M., and Stanford, C. B. 2008. *Beautiful minds: The parallel lives of great apes and dolphins*. Cambridge, MA: Harvard University Press.

Betzig, L. L. 1986. *Despotism and differential reproduction: A Darwinian view of history*. New York: Aldine.

Beyene, Y. 2010. Herto brains and minds: Behaviour of early *Homo sapiens* from the middle awash, Ethiopia. In *Social brain, distributed mind,* eds. R. Dunbar, C. Gamble, and J. Gowlett. New York: Oxford University Press.

Binford, L. 1978. *Nunamiut ethnoarchaeology*. New York: Academic Press.

——. 2001. *Constructing frames of reference: An analytical method for archaeological theory building using hunter–gatherer and environmental data sets*. Berkeley and Los Angeles: University of California Press.

Bird, R. B., Smith, E. A., and Bird, D. W. 2001. The hunting handicap: Costly signaling in human foraging strategies. *Behavioral Ecology and Sociobiology* 50:9–19.

Bird-David, N. 1992. Beyond "The original affluent society": A culturalist reformation. *Current Anthropology* 33:25–48.

Black, D. 2011. *Moral time*. New York: Oxford University Press.

Blurton Jones, N. G. 1991. Tolerated theft: Suggestions about the ecology and evolution of sharing, hoarding, and scrounging. In *Primate politics,* eds. G. Schubert and R. D. Masters. Carbondale: Southern Illinois University Press.

Boehm, C. 1972. Montenegrin ethical values: An experiment in anthropological method. PhD diss., Harvard University.

——. 1976. Biological versus social evolution. *American Psychologist* 31:348–351.

——. 1978. Rational pre-selection from Hamadryas to *Homo sapiens:* The place of decisions in adaptive process. *American Anthropologist* 80:265–296.

——. 1979. Some problems with "altruism" in the search for moral universals. *Behavioral Science* 24:15–24.

——. 1980. Exposing the moral self in Montenegro: The use of natural definitions in keeping ethnography descriptive. *American Ethnologist* 7:1–26.

——. 1981. Parasitic selection and group selection: A study of conflict interference in Rhesus and Japanese Macaque monkeys. In *Primate behavior and sociobiology: Proceedings of the international congress of primatology,* eds. A. B. Chiarelli and R. S. Corruccini. Heidelberg, Germany: Springer.

——. 1982. The evolutionary development of morality as an effect of dominance behavior and conflict interference. *Journal of Social and Biological Sciences* 5:413–422.

——. 1983. *Montenegrin social organization and values*. New York: AMS Press.

———. 1985. Execution within the clan as an extreme form of ostracism. *Social Science Information* 24:309–321.

———. 1986. *Blood revenge: The enactment and management of conflict in Montenegro and other tribal societies*. Philadelphia: University of Pennsylvania Press.

———. 1989. Ambivalence and compromise in human nature. *American Anthropologist* 91:921–939.

———. 1991a. Lower-level teleology in biological evolution: Decision behavior and reproductive success in two species. *Cultural Dynamics* 4:115–134.

———. 1991b. Response to Knauft, violence and sociality in human evolution. *Current Anthropology* 32:411–412.

———. 1993. Egalitarian behavior and reverse dominance hierarchy. *Current Anthropology* 34:227–254.

———. 1996. Emergency decisions, cultural selection mechanics, and group selection. *Current Anthropology* 37:763–793.

———. 1997. Impact of the human egalitarian syndrome on Darwinian selection mechanics. *American Naturalist* 150:100–121.

———. 1999. *Hierarchy in the forest: The evolution of egalitarian behavior*. Cambridge, MA: Harvard University Press.

———. 2000. Conflict and the evolution of social control. *Journal of Consciousness Studies*, Special Issue on Evolutionary Origins of Morality. L. Katz, ed. 7:79–183.

———. 2002. Variance reduction and the evolution of social control. Paper presented at Santa Fe Institute, Fifth Annual Workshop on the Co-evolution of Behaviors and Institutions, Santa Fe, New Mexico. www.santafe.edu/files/gems/behavioralsciences/variance.pdf.

———. 2003. Global conflict resolution: An anthropological diagnosis of problems with world governance. In *Evolutionary psychology and violence: A primer for policymakers and public policy advocates*, eds. R. W. Bloom and N. Dess. London: Praeger.

———. 2004a. Explaining the prosocial side of moral communities. In *Evolution and ethics: Human morality in biological and religious perspective*, eds. P. Clayton and J. Schloss. New York: Eerdmans.

———. 2004b. What makes humans economically distinctive? A three-species evolutionary comparison and historical analysis. *Journal of Bioeconomics* 6:109–135.

———. 2007. The natural history of blood revenge. In *Feud in medieval and early modern Europe*, eds. B. Poulsen and J. B. Netterström. Aarhus, Denmark: Aarhus University Press.

——. 2008a. A biocultural evolutionary exploration of supernatural sanctioning. In *Evolution of religion: Studies, theories, and critiques,* eds. J. Bulbulia, R. Sosis, R. Genet, E. Harris, K. Wyman, and C. Genet. Santa Margarita, CA: Collins Family Foundation.

——. 2008b. Purposive social selection and the evolution of human altruism. *Cross-Cultural Research* 42:319–352.

——. 2009. How the golden rule can lead to reproductive success: A new selection basis for Alexander's "indirect reciprocity." In *The golden rule: Analytical perspectives,* eds. J. Neusner and B. Chilton. Lanham, MD: University Press of America.

——. 2011. Retaliatory violence in human prehistory. *British Journal of Criminology* 51:518–534.

Boehm, C. 2012. Variance reduction and the evolution of social control: A methodology for the reconstruction of ancestral social behavior from evidence on ethnographic foragers. Working papers, Santa Fe Institute.

Boehm, C. In press. Costs and benefits in hunter-gatherer punishment. Commentary on Francisco Guala, Reciprocity: Weak or strong? What punishment experiments do and do not demonstrate. *Behavioral and Brain Sciences.*

Boehm, C., and Flack, J. 2010. The emergence of simple and complex power structures through social niche construction. In *The social psychology of power,* ed. A. Guinote. New York: Guilford Press.

Boesch, C. 1991. The effects of leopard predation on grouping patterns in forest chimpanzees. *Behaviour* 117:220–241.

Boesch, C., and Boesch-Achermann, H. 1991. Dim forest, bright chimps. *Natural History* 9:50–56.

——. 2000. *The chimpanzees of the Taï forest: Behavioural ecology and evolution.* New York: Oxford University Press.

Bogardus, E. S. 1933. A social distance scale. *Sociology and Social Research* 17:265–271.

Bowles, S. 2006. Group competition, reproductive leveling, and the evolution of human altruism. *Science* 314:1569–1572.

——. Did warfare among ancestral hunter-gatherers affect the evolution of human social behaviors? *Science* 324:1293–1298.

Bowles, S., and Gintis, H. 2004. The evolution of strong reciprocity: Cooperation in heterogeneous populations. *Theoretical Population Biology* 65:17–28.

——. 2011. *A cooperative species: Human reciprocity and its evolution.* Princeton, NJ: Princeton University Press.

Boyd, R., Gintis, H., Bowles, S., and Richerson, P. J. 2003. The evolution of altruistic punishment. *Proceedings of the National Academy of Sciences* 100:3531–3535.

Boyd, R., and Richerson, P. J. 1985. *Culture and the evolutionary process.* Chicago: University of Chicago Press.

——. 1992. Punishment allows the evolution of cooperation or anything else in sizable groups. *Ethology and Sociobiology* 13:171–195.

Breasted, J. H. 1933. *The dawn of conscience.* New York: Scribner's.

Briggs, J. L. 1970. *Never in anger: Portrait of an Eskimo family.* Cambridge, MA: Harvard University Press.

——. 1982. Living dangerously: The contradictory foundations of value in Canadian Inuit society. In *Politics and history in band societies,* eds. E. Leacock and R. Lee. Cambridge: Cambridge University Press.

——. 1994. "Why don't you kill your baby brother?" The dynamics of peace in Canadian Inuit camps. In *The anthropology of peace and nonviolence,* eds. L. E. Sponsel and T. Gregor. Boulder, CO: Lynne Rienner.

——. 1998. *Inuit morality play: The emotional education of a three-year-old.* New Haven, CT: Yale University Press.

Brosnan, S. F. 2006. Nonhuman species' reactions to inequity and their implications for fairness. *Social Justice Research* 19:153–185.

Brown, D. 1991. *Human universals*. New York: McGraw-Hill.

Brown, S. L., Nesse, R. M., Vinokur, A. D., and Smith, D. M. 2003. Providing social support may be more beneficial than receiving it: Results from a prospective study of mortality. *Psychological Science* 14:320–327.

Bunn, H. T., and Ezzo, J. A. 1993. Hunting and scavenging by Plio-Pleistocene hominids: Nutritional constraints, archaeological patterns, and behavioural implications. *Journal of Archaeological Science* 20:365–398.

Burch Jr., E. S. 2005. *Alliance and conflict: The world system of the* Iñupiaq *Eskimos*. Lincoln: University of Nebraska Press.

Burroughs, W. J. 2005. *Climate change in prehistory: The end of the reign of chaos*. Cambridge: Cambridge University Press.

Byrne, R. W. 1993. Empathy in primate social manipulation and communication: A precursor to ethical behaviour. In *Biological evolution and the emergence of ethical conduct,* ed. G. Thines, Bruxelles: Académie Royale de Belgique.

Byrne, R. W., and J. M. Byrne 1988. Leopard killers of Mahale. *Natural History* 97:22–26.

Campbell, D. T. 1965. Variation and selective retention in socio-cultural evolution. In *Social change in developing areas,* eds. H. R. Barringer, B. I. Blanksten, and R. W. Mack. Cambridge, MA: Schenkman.

——. 1972. On the genetics of altruism and the counter-hedonic component of human culture. *Journal of Social Issues* 28:21–37.

——. 1975. On the conflicts between biological and social evolution and between psychology and moral tradition. *American Psychologist* 30:1103–1126.

Cantrell, P. J. 1994. Family violence and incest in Appalachia. *Journal of the Appalachian Studies Association* 6:39–47.

Casimir, M. J., and Schnegg, M. 2002. Shame across cultures: The evolution, ontogeny, and function of a "moral emotion." In *Between culture and biology: Perspectives on ontogenetic development*, eds. H. Keller, Y. H. Poortinga, and A. Scholmerich. Cambridge: Cambridge University Press.

Cavalli-Sforza, L. L., and Edwards, A. W. F. 1967. Phylogenetic analysis: Models and estimation procedures. *Evolution* 32:550–570.

Choi, J.-K., and Bowles, S. 2007. The coevolution of parochial altruism and war. *Science* 26:636–640.

Coser, L. 1956. *The functions of social conflict*. New York: Free Press.

Cosmides, L., Tooby, J., Fiddick, L., and Bryant, G. A. 2005. Detecting cheaters. *Trends in Cognitive Sciences* 9:505–506.

Cummins, D. D. 1999. Cheater detection is modified by social rank: The impact of dominance on the evolution of cognitive functions. *Evolution and Human Behavior* 20:229–248.

Damasio, A. R. 2002. The neural basis of social behavior: Ethical implications. Paper presented at the conference Neuroethics: Mapping the Field, San Francisco, California, May 13–14.

Damasio, H., Grabowski, T., Frank, R., Galaburda, A. M., and Damasio, A. R. 1994. The return of Phineas Gage: Clues about the brain from the skull of a famous patient. *Science* 264:1102–1105.

Darwin, C. 1859. *On the origin of species*. London: John Murray.

——. 1865. *The expression of the emotions in man and animals*. Chicago: University of Chicago Press.

——. 1982 (1871). *The descent of man, and selection in relation to sex*. Princeton, NJ: Princeton University Press.

Dawkins, R. 1976. *The selfish gene*. New York: Oxford University Press.

Dewey, J. 1934. *Art as experience*. New York: Minton, Balch.

Diamond, J. 1992. *The third chimpanzee: The evolution and future of the human animal*. New York: Harper Perennial.

Draper, P. 1978. The learning environment for aggression and anti-social behavior among the !Kung. In *Learning non-aggression: The experience of nonliterate societies*, ed. A. Montagu. New York: Oxford University Press.

Dubreuil, B. 2010. Paleolithic public goods games: Why human culture and cooperation did not evolve in one step. *Biology and Philosophy* 25:53–73.

Dunbar, R. I. M. 1996. *Grooming, gossip, and the evolution of language*. London: Faber and Faber.

Durham, W. H. 1991. *Coevolution: Genes, culture, and human diversity*. Stanford, CA: Stanford University Press.

Durkheim, É. 1933. *The division of labor in society*. New York: Free Press.

Dyson-Hudson, R., and Smith, E. A. 1978. Human territoriality: An ecological reassessment. *American Anthropologist* 80:21–41.

Eibl-Eibesfeldt, I. 1982. Warfare, man's indoctrinability, and group selection. *Zeitschrift für Tierpsychologie* 60:177–198.

Eisenberg, N., Fabes, R. A. and Spinrad, T. L. 2006. Prosocial development. In, *Handbook of child psychology, Volume 3: Social, personal, and personality development*, ed. N. Eisenberg, New York: Wiley.

Eldredge, N. 1971. The allopatric model and phylogeny in Paleozoic invertebrates. *Evolution* 25:156–167.

Elkin, A. P. 1994. *Aboriginal men of high degree: Initiation and sorcery in the world's oldest tradition*. Rochester, VT: Inner Traditions.

Ellis, L. 1995. Dominance and reproductive success among nonhuman animals: A cross-species comparison. *Ethology and Sociobiology* 16:257–333.

Erdal, D., and Whiten, A. 1994. On human egalitarianism: An evolutionary product of Machiavellian status escalation? *Current Anthropology* 35:175–184.

Faulkner, W. 1954. *A fable*. New York: Random House.

Fehr, E. 2004. Don't lose your reputation. *Nature* 432:449–450.

Fehr, E., Bernhard, H. and Rockenbach, B. 2008. Egalitarianism in young children. *Nature* 454:1079–1084.

Fehr, E., and Gächter, S. 2002. Altruistic punishment in humans. *Nature* 415:137–140.

——. 2004. Reply to Fowler et al.: Egalitarian motive and altruistic punishment. *Nature* 433:E1–E2.

Fisher, R. A. 1930. *The genetical theory of natural selection*. New York: Dover.

Flack, J. C., and de Waal, F. B. M. 2000. "Any animal whatever": Darwinian building blocks of morality in monkeys and apes. *Journal of Consciousness Studies* 7:1–29.

Fleagle, J. G., and Gilbert, C. C. 2008. Modern human origins in Africa. *Evolutionary Anthropology* 17:1–2.

Fouts, R., with Mills, S. T. 1997. *Next of kin: My conversations with chimpanzees*. New York: Avon.

Frank, R. H. 1988. *Passions within reason: The strategic role of the emotions.* New York: Norton.

Frank, S. A. 1995. Mutual policing and repression of competition in the evolution of cooperative groups. *Nature* 377:520–522.

Freud, S. 1918. *Totem and taboo: Resemblances between the psychic lives of savages and neurotics.* Trans. A. A. Brill. New York: Random House.

Fry, D. P. 2000. Conflict management in cross-cultural perspective. In *Natural conflict resolution,* eds. F. Aureli and F. B. M. de Waal. Berkeley and Los Angeles: University of California Press.

Furíuchi, Takeshi 2011. Female contributions to the peaceful nature of bonobo society. *Evolutionary Anthropology* 20:131–142.

Gallup, G. G. J., Anderson, J. R., and Shillito, D. J. 2002. The mirror test. In *The cognitive animal: Empirical and theoretical perspectives on animal cognition,* eds. M. Bekoff, C. Allen, and G. Burghardt. Cambridge, MA: MIT Press.

Gardner, B. T., and Gardner, R. A. 1994. Development of phrases in the utterances of children and cross-fostered chimpanzees. In *The ethological roots of culture,* eds. R. A. Gardner, B. T. Gardner, B. Chiarelli, and F. X. Plooj. London: Kluwer Academic.

Ghiselin, M. T. 1974. *The economy of nature and the evolution of sex.* Berkeley and Los Angeles: University of California Press.

Gintis, H. 2003. The hitchhiker's guide to altruism: Gene-culture coevolution and the internalization of norms. *Journal of Theoretical Biology* 220:407–418.

Goodall, J. 1986. *The chimpanzees of Gombe: Patterns of behavior.* Cambridge, MA: Belknap Press.

——. 1992. Unusual violence in the overthrow of an alpha male chimpanzee at Gombe. In *Topics in primatology,* vol. 1: *Human origins,* eds. T. Nishida, W. C. McGrew, P. Marler, M. Pickford, and F. B. M. de Waal. Tokyo: University of Tokyo Press.

Gould, R. A. 1982. To have and have not: The ecology of sharing among hunter-gatherers. In *Resource managers: North American and Australian hunter-gatherers,* eds. N. M. Williams and E. S. Hunn. Boulder, CO: Westview Press.

Greene, J. D. 2003. From neural "is" to moral "ought": What are the moral implications of neuroscientific moral psychology? *Nature Reviews Neuroscience* 4:847–850.

——. 2007. The secret joke of Kant's soul. In *Moral psychology,* vol. 3: *The neuroscience of morality,* ed. W. Sinnott-Armstrong. Cambridge, MA: MIT Press.

Guala, F. In press. Reciprocity: Weak or strong? What punishment experiments do and do not demonstrate. *Behavioral and Brain Sciences.*

Gurven, M. 2004. To give and give not: The behavioral ecology of human food transfers. *Behavioral and Brain Sciences* 27:543–583.

Gurven, M., Allen-Arave, W., Hill, K., and Hurtado, A. M. 2000. "It's a wonderful life": Signaling generosity among the Ache of Paraguay. *Evolution and Human Behavior* 21:263–282.

——. 2001. Reservation food sharing among the Ache of Paraguay. *Human Nature* 12:273–297.

Gurven, M., and Hill, K. 2010. Moving beyond stereotypes of men's foraging goals. *Current Anthropology* 51:265–267.

Haidt, J. 2003. The moral emotions. In *Handbook of affective sciences,* eds. R. J. Davidson, K. R. Scherer, and H. H. Goldsmith. New York: Oxford University Press.

——. 2007. The new synthesis in moral psychology. *Science* 18:998–1002.

Haile, B. 1978. *Love-magic and the butterfly people: The Slim Curly version of the Ajilee and Mothway myths.* Flagstaff: Museum of Arizona Press.

Hamilton, W. D. 1964. The genetical evolution of social behavior I, II. *Journal of Theoretical Biology* 7:1–52.

——. 1975. Innate social aptitudes in man: An approach from evolutionary genetics. In *Biosocial anthropology,* ed. R. Fox. London: Malaby.

Hammerstein, P., and Hagen, E. H. 2004. The second wave of evolutionary economics in biology. *Trends in Ecology and Evolution* 20:604–609.

Hammerstein, P., and Hoekstra, R. F. 2002. Mutualism on the move. *Nature* 376:121–122.

Hare, R. 1993. *Without conscience: The disturbing world of the psychopaths among us.* New York: Guilford Press.

Harpending, H., and Rogers, A. 2000. Genetic perspectives on human origins and differentiation. *Annual Review of Genomics and Human Genetics* 1:361–385.

Hauser, M. D. 2006. *Moral minds: How nature designed our universal sense of right and wrong.* New York: HarperCollins.

Haviland, J. B. 1977. *Gossip, reputation, and knowledge in Zinacantan.* Chicago: University of Chicago Press.

Hawkes, K. 1991. Showing off: Tests of an hypothesis about men's foraging goals. *Ethology and Sociobiology* 12:29–54.

——. 2001. Is meat the hunter's property? Big game, ownership, and explanations of hunting and sharing. In *Meat-eating and human evolution,* eds. C. B. Stanford and H. T. Bunn. New York: Oxford University Press.

Hawks, J., Hunley, K., Lee, S.-H., and Wolpoff, M. 2000. Population bottlenecks and Pleistocene human evolution. *Molecular Biology and Evolution* 17:2–22.

Heinz, H. J. 1994. *Social organization of the !Ko Bushmen*. Cologne, Germany: Rüdiger Köppe Verlag.

Henrich, J., Boyd, R., Bowles, S., Camerer, C., Fehr, E., Gintis, H., McElreath, R., Alvard, M., Barr, A., Ensminger, J., Hill, K., Gil-White, F., Gurven, M., Marlowe, F., Patton, J. Q., Smith, N., and Tracer, D. 2005. "Economic man" in cross-cultural perspective: Behavioral experiments in 15 small-scale societies. *Behavioral and Brain Sciences* 28:795–855.

Hewlett, B. S., and Lamb, M. E., eds. 2006. *Hunter-gatherer childhoods: Evolutionary, developmental, and cultural perspectives*. New Brunswick, NJ: Transaction.

Hill, K. R., Walker, R., Boievi, M., Eder, J., Headland, T., Hewlett, B., Hurtado, A. M., Marlowe, F., Wiessner, P., and Wood, B. 2011. Coresidence patterns in hunter-gatherer societies show unique human social structure. *Science* 331:1286–1289.

Hohmann, G., and Fruth, B. 1993. Field observations on meat sharing among bonobos. *Folia Primatologica* 60:225–229.

Hrdy, S. B. 2009. *Mothers and others: The evolutionary origins of mutual understanding*. Cambridge, MA: Belknap Press.

Huxley, T. H. 1894. *Evolution and ethics*. New York: Appleton.

Irons, W. 1991. How did morality evolve? *Zygon* 26:49–89.

Johnson, D. D. P., and Krüger, O. 2004. The good of wrath: Supernatural punishment and the evolution of cooperation. *Political Theology* 5:159–176.

Jones, D. M. 1969. A study of social and economic problems in Unalaska, an Aleut village. PhD diss., University of California Berkeley. University microfilms, publications 70–6048.

Kagan, J. 1981. *The second year: The emergence of self-awareness*. Cambridge, MA: Harvard University Press.

Kagan, J., and Lamb, S., eds. 1987. *The emergence of morality in young children*. Chicago: University of Chicago Press.

Kano, T. 1992. *The last ape: Pygmy chimpanzee behavior and ecology*. Stanford, CA: Stanford University Press.

Kaplan, H., and Gurven, M. 2005. The natural history of human food sharing and cooperation: A review and a new multi-individual approach to the negotiation of norms. In *Moral sentiments and material interests: On the foundations of cooperation in economic life,* eds. H. Gintis, S. Bowles, R. Boyd, and E. Fehr. Cambridge, MA: MIT Press.

Kaplan, H., and Hill, K. 1985. Food sharing among Aché foragers: Tests of explanatory hypotheses. *Current Anthropology* 26:223–246.

Katz, L. D., ed. 2000. *Evolutionary origins of morality: Cross-disciplinary perspectives*. Bowling Green, OH: Imprint Academic.

Keeley, L. 1988. Hunter-gatherer economic complexity and "population pressure": A cross-cultural analysis. *Journal of Anthropological Archaeology* 7:373–411.

——. 1996. *War before civilization*. New York: Oxford University Press.

Keenan, J. P., Gallup Jr., G. G., and Falk, D. 2003. *The face in the mirror: The search for the origins of consciousness*. New York: HarperCollins.

Kelly, R. C. 2000. *Warless societies and the evolution of war*. Ann Arbor: University of Michigan Press.

——. 2005. The evolution of lethal intergroup violence. *Proceedings of the National Academy of Sciences* 102:15294–15298.

Kelly, R. L. 1995. *The foraging spectrum: Diversity in hunter-gatherer lifeways*. Washington, DC: Smithsonian Institution Press.

Kiehl, K. A. 2008. Without morals: The cognitive neuroscience of criminal psychopaths. In *Moral psychology*, vol. 1: *The evolution of morality: Adaptations and innateness*, ed. W. Sinnott-Armstrong. Cambridge, MA: MIT Press.

Kiehl, K. A., Bates, A. T., Laurens, K. R., Hare, R. D., and Liddle, P. F. 2006. Brain potentials implicate temporal lobe abnormalities in criminal psychopaths. *Journal of Abnormal Psychology* 115:443–453.

Klein, R. G. 1999. *The human career: Human biological and cultural origins*. Chicago: University of Chicago Press.

Knauft, B. M. 1991. Violence and sociality in human evolution. *Current Anthropology* 32:391–428.

Kollock, P. 1998. Social dilemmas: The anatomy of cooperation. *Annual Review of Sociology* 24:183–214.

Konner, M. 2010. *The evolution of childhood: Relationships, emotion, mind*. Cambridge, MA: Harvard University Press.

Krebs, D. L. 2000. The evolution of moral dispositions in the human species. *Annals of the New York Academy of Sciences* 907:132–148.

Ladd, C., and Maloney, K. 2011. Chimp murder at Mahale. http://www.nomad-tanzania.com/blogs/greystoke-mahale/murder-in-mahale.

Laland, K. N., Odling-Smee, J., and Feldman, M. W. 2000. Niche construction, biological evolution, and cultural change. *Behavioral and Brain Sciences* 23:131–175.

Laland, K. N., Odling-Smee, J., and Myles, S. 2010. How culture shaped the human genome: Bringing genetics and the human sciences together. *Nature Reviews Genetics* 11:137–148.

Laughlin, C. D., and Brady, I. A., eds. 1978. *Extinction and survival in human populations*. New York: Columbia University Press.

Leacock, E. 1969. The Montagnais-Naskapi band. In *Contributions to anthropology: Band societies*, ed. D. Damas. Bulletin 228. Ottawa, ON: National Museum of Canada.

Lee, R. B. 1979. *The !Kung San: Men, women, and work in a foraging society*. Cambridge: Cambridge University Press.

LeVine, R. A., and Campbell, D. T. 1972. *Ethnocentrism: Theories of conflict, ethnic attitudes, and group behavior*. New York: Wiley.

Lewontin, R. C. 1970. The units of selection. *Annual Review of Ecology and Systematics* 1:1–18.

Lindsay, S. 2000. *Handbook of applied dog behavior and training*, vol. 1: *Adaptation and learning*. Ames: Iowa State University Press.

Lore, R., Nikoletseas, M., and Takahashi, L. 1984. Colony aggression in laboratory rats: A review and some recommendations. *Aggressive Behavior* 10:59–71.

Lorenz, K. 1966. *On aggression*. New York: Bantam.

Lyell, C. 1833. *Principles of geology, being an attempt to explain the former changes of the Earth's surface, by reference to causes now in operation*. Vol. 3. London: John Murray.

Malinowski, B. 1922. *Argonauts of the western Pacific: An account of native enterprise and adventure in the archipelagoes of Melanesian New Guinea*. New York: Dutton.

——. 1929. *The sexual life of savages in northwestern Melanesia*. London: George Routledge.

Malthus, T. R. 1985 (1798). *An essay on the principle of population*. New York: Penguin.

Marlowe, F. W. 2004. Mate preferences among Hadza hunter-gatherers. *Human Nature* 15:365–376.

——. 2005. Hunter-gatherers and human evolution. *Evolutionary Anthropology* 14:54–67.

——. 2010. *The Hadza hunter-gatherers of Tanzania*. Berkeley and Los Angeles: University of California Press.

Mayr, E. 1983. How to carry out the adaptationist program? *American Naturalist* 121:324–334.

——. 1988. The multiple meanings of teleological. In *Towards a new philosophy of biology*, ed. E. Mayr. Cambridge, MA: Harvard University Press.

——. 1997. *This is biology*. Cambridge, MA: Harvard University Press.

——. 2001. *What evolution is*. New York: Basic Books.

McBrearty, S., and Brooks, A. 2000. The revolution that wasn't: A new interpretation of the origin of modern human behavior. *Journal of Human Evolution* 39:453–563.

McCullough, M. E., Kimeldorf, M. B., and Cohen, A. D. 2008. An adaptation for altruism? The social causes, social effects, and social evolution of gratitude. *Current Directions in Psychological Science* 17:281–285.

Mead, G. H. 1934. *Mind, self, and society*. Chicago: University of Chicago Press.

Menzel, E. W. 1974. A group of young chimpanzees in a one acre field. In *Behavior of non-human primates*. Vol. 5, eds. A. M. Shrier and F. Stollnitz. New York: Academic Press.

Midgley, M. 1994. *The ethical primate: Humans, freedom, and morality*. London: Routledge.

Mirsky, J. 1937. The Eskimo of Greenland. In *Cooperation and competition among primitive peoples*, ed. M. Mead. New York: McGraw-Hill.

Mithen, S. J. 1990. *Thoughtful foragers: A study of prehistoric decision making*. Cambridge: Cambridge University Press.

Myers, F. R. 1988. Burning the truck and holding the country: Property, time, and the negotiation of identity among Pintupi Aborigines. In *Hunters and gatherers*, vol. 2: *Property, power, and ideology*, eds. T. Ingold, D. Riches, and J. Woodburn. Oxford: Berg.

Nesse, R. M. 2000. How selfish genes shape moral passions. *Journal of Consciousness Studies* 7:227–231.

——. 2007. Runaway social selection for displays of partner value and altruism. *Biological Theory* 2:143–155.

——. 2010. Social selection and the origins of culture. In *Evolution, culture, and the human mind*, eds. M. Schaller, A. Norenzayan, S. J. Heine, T. Yamagishi, and T. Kameda. Philadelphia: Erlbaum.

Neusner, J., and Chilton, B., eds. 2009. *The golden rule: Analytical perspectives*. Lanham, MD: University Press of America.

Nietzsche, F. 1887. *Zur genealogie der moral: Eine streitschrift* (On the genealogy of morals: A polemical tract). Leipzig, Germany: Verlag von C. G. Naumann.

Nishida, T. 1996. The death of *Ntologi*, the unparalleled leader of M group. *Pan Africa News* 3:4.

Noss, A. J., and Hewlett, B. S. 2001. The contexts of female hunting in central Africa. *American Anthropologist* 103:1024–1040.

Nowak, M. A., and Sigmund, K. 2005. Evolution of indirect reciprocity. *Nature* 437:1291–1298.

Nowak, M. A., Tarnita, C. E., and Wilson, E. O. 2010. The evolution of eusociality. *Nature* 466:1057–1062.

Otterbein, K. F. 1988. Capital punishment: A selection mechanism. Commentary on Robert K. Dentan, On Semai homicide. *Current Anthropology* 29:633–636.

Panchanathan, K., and Boyd, R. 2004. Indirect reciprocity can stabilize cooperation without the second-order free rider problem. *Nature* 432:499–502.

Parker, I. 2007. Swingers: Bonobos are celebrated as peace-loving, matriarchal, and sexually liberated. Are they? *New Yorker,* July, 48–61.

Parsons, T., and Shils, E., eds. 1952. *Toward a general theory of action*. Cambridge, MA: Harvard University Press.

Patterson, F., and Linden, E. 1981. *The education of Koko*. New York: Holt, Reinhart and Winston.

Pericot, L. 1961. The social life of Spanish Paleolithic hunters as shown by Levantine art. In *Social life of early man,* ed. S. L. Washburn. Chicago: Aldine.

Peterson, N. 1993. Demand sharing: Reciprocity and the pressure for generosity among foragers. *American Anthropologist* 95:860–874.

Piers, G., and Singer, M. B. 1971. *Shame and guilt: A psychoanalytic and a cultural study*. New York: Norton.

Pinker, S. 2011. *The better angels of our nature: Why violence has declined.* New York: Viking.

Popper, K. 1978. Natural selection and the emergence of mind. *Dialectica* 32:339–355.

Potts, R. 1996. *Humanity's descent: The consequences of ecological instability.* New York: Aldine de Gruyter.

Preston, S. D., and de Waal, F. B. M. 2002. Empathy: Its ultimate and proximate bases. *Behavioral and Brain Sciences* 25:1–72.

Price, M. E., Cosmides, L., and Tooby, J. 2002. Punitive sentiment as an anti–free rider psychological device. *Evolution and Human Behavior* 23:203–231.

Rapoport, A., and Chammah, A. 1965. *Prisoner's dilemma*. Ann Arbor: University of Michigan Press.

Rasmussen, K. 1931. *The Netsilik Eskimos: Social life and spiritual culture. Report of the fifth Thule Expedition 1921–24,* Vol. VIII, No. 1–2. Copenhagen: Gyldendalske Boghandel.

Richards, R. J. 1989. *Darwin and the emergence of evolutionary theories of mind and behavior*. Chicago: University of Chicago Press.

Richerson, P. J., and Boyd, R. 1999. Complex societies: The evolutionary origins of a crude superorganism. *Human Nature* 10:253–289.

Riches, D. 1974. The Netsilik Eskimo: A special case of selective female infanticide. *Ethnology* 13:351–361.

Ridley, M. 1996. *The origins of virtue: Human instincts and the evolution of cooperation*. New York: Penguin.

Ruvolo, M., Disotell, T. R., Allard, M. W., Brown, W. M., and Honeycutt, R. L. 1991. Resolution of the African hominoid trichotomy by use of a mitochondrial gene sequence. *Proceedings of the National Academy of Science* 88:1570–1574.

Savage-Rumbaugh, S., and Lewin, R. 1994. *Kanzi: The ape at the brink of the human mind*. New York: Wiley.

Savage-Rumbaugh, S., Shanker, S. G., and Taylor, T. J. 1998. *Apes, languages, and the human mind*. New York: Oxford University Press.

Service, E. R. 1975. *Origin of the state and civilization: The process of cultural evolution*. New York: Norton.

Sherman, P. W., Alexander, R. D., and Jarvis, J. U. 1991. *The biology of the naked mole rat*. Princeton, NJ: Princeton University Press.

Shermer, M. 2004. *The science of good and evil: Why people cheat, gossip, care, share, and follow the golden rule*. New York: Henry Holt.

Shostak, M. 1981. *Nisa: The life and words of a !Kung woman*. Cambridge, MA: Harvard University Press.

Silberbauer, G. B. 1981. *Hunter and habitat in the central Kalahari desert*. Cambridge: Cambridge University Press.

Simon, H. A. 1990. A mechanism for social selection and successful altruism. *Science* 250:1665–1668.

Skyrms, B. 2000. Game theory, rationality, and evolution of the social contract. In *Evolutionary origins of morality: Cross-disciplinary perspectives*, ed. L. D. Katz. Bowling Green, OH: Imprint Academic.

Smith, E. A. 2004. Why do good hunters have higher reproductive success? *Human Nature* 15:343–364.

Smith, E. A., and Boyd, R. 1990. Risk and reciprocity: Hunter-gatherer socioecology and the problem of collective action. In *Risk and uncertainty in tribal and peasant economies*, ed. E. A. Cashdan. Boulder, CO: Westview Press.

Sober, E., and Wilson, D. S. 1998. *Unto others: The evolution and psychology of unselfish behavior*. Cambridge, MA: Harvard University Press.

——. 2000. Summary of *Unto others: The evolution and psychology of unselfish behavior*. In *Evolutionary origins of morality: Cross-disciplinary perspectives*, ed. L. D. Katz. Bowling Green, OH: Imprint Academic.

Spencer, H. 1851. *Social statistics; or the conditions essential to human happiness specified, and the first of them developed.* London: John Chapman.

Speth, J. D. 1989. Early hominid hunting and scavenging: The role of meat as an energy source. *Journal of Human Evolution* 18:329–343.

Sprengel, K. P. 1839. *Die lehre vom dünger oder beschreibung aller bei der landwirthschaft gebräuchlicher vegetabilischer, animalischer und mineralischer düngermaterialien, nebst erklärung ihrer wirkungsart* (Principles of fertilization in a description of the vegetable, animal, and mineral fertilizers employed in agriculture with an explanation of their mode of action). Leipzig, Germany: Verlag.

Stanford, C. B. 1999. *The hunting apes: Meat eating and the origins of human behavior.* Princeton, NJ: Princeton University Press.

Stephenson, J. 2000. *The language of the land: Living among the Hadzabe in Africa.* New York: St. Martin's Press.

Stevens, J. R., Cushman, F. A., and Hauser, M. D. 2005. Evolving the psychological mechanisms for cooperation. *Annual Review of Ecology, Evolution, and Systematics* 36:499–518.

Steward, J. H. 1938. Basin-plateau Aboriginal sociopolitical groups. Smithsonian Institution Bureau of American Ethnology Bulletin 120. Washington, DC: GPO.

——. 1955. *Theory of culture change.* Urbana: University of Illinois Press.

Stiner, M. C. 2002. Carnivory, coevolution, and the geographic spread of the genus *homo. Journal of Archaeological Research* 10:1–63.

Stiner, M. C., Barkai, R., and Gopher, A. 2009. Cooperative hunting and meat sharing 400–200 kya at Qesem cave, Israel. *Proceedings of the National Academy of Sciences* 106:13207–13212.

Sullivan, R. J. 1989. *Immanuel Kant's moral theory.* Cambridge: Cambridge University Press.

Temerlin, M. K. 1975. *Lucy: Growing up human: A chimpanzee daughter in a psychotherapist's family.* Palo Alto, CA: Science and Behavior Books.

Testart, A. 1982. The significance of food storage among hunter-gatherers: Residence patterns, population densities, and social inequalities. *Current Anthropology* 23:523–537.

Thieme, H. 1997. Lower Paleolithic hunting spears from Germany. *Nature* 385:807.

Thomas, E. M. 1989. *The harmless people.* New York: Vintage.

Tiger, L. 1979. *Optimism: The biology of hope.* New York: Simon and Schuster.

Trivers, R. L. 1971. The evolution of reciprocal altruism. *Quarterly Review of Biology* 46:35–57.

——. 1972. Parental investment and sexual selection. In *Sexual selection and the descent of man, 1871–1971,* ed. B. G. Campbell. Chicago: Aldine.

Trut, L., Oskina, I., and Kharlamova, A. 2009. Animal evolution during domestication: The domesticated fox as a model. *BioEssays* 31:349–360.

Turiel, Eliot 2006. The development of morality. In, *Handbook of child psychology, Volume 3: Social, personal, and personality development,* ed. N. Eisenberg, New York: Wiley.

Turnbull, C. M. 1961. *The forest people.* Garden City, NY: Natural History Press.

——. 1972. *The mountain people.* New York: Simon and Schuster.

van den Steenhoven, G. 1957. *Research report on Caribou Eskimo law.* The Hague, the Netherlands: G. van den Steenhoven.

——. 1959. Legal concepts among the Netsilik Eskimos of Pelly Bay, Northwest Territories. NCRC Report 59–3. Ottawa, ON: Canada Department of Northern Affairs.

——. 1962. *Leadership and law among the Eskimos of the Keewatin district, Northwest Territories.* Rijswijk, the Netherlands: Excelsior.

Voland, E., and Voland, R. 1995. Parent-offspring conflict, the extended phenotype, and the evolution of conscience. *Journal of Social and Evolutionary Systems* 18:397–412.

von Furer-Haimendorf, C. 1967. *Morals and merit: A study of values and social controls in South Asian societies.* Chicago: University of Chicago Press.

de Waal, F. B. M. 1982. *Chimpanzee politics: Power and sex among apes.* New York: Harper and Row.

——. 1986. The brutal elimination of a rival among captive male chimpanzees. *Ethology and Sociobiology* 7:237–251.

——. 1996. *Good natured: The origins of right and wrong in humans and other animals.* Cambridge, MA: Harvard University Press.

——. 2008. Putting the altruism back into altruism: The evolution of empathy. *Annual Review of Psychology* 59:279–300.

——. 2009. *The age of empathy: Nature's lessons for a kinder society.* New York: Harmony Books.

de Waal, F. B. M., and Lanting, F. 1997. *Bonobo: The forgotten ape.* Berkeley and Los Angeles: University of California Press.

Waddington, C. H. 1960. *The ethical animal.* Chicago: University of Chicago Press.

Wade, M. J. 1978. A critical review of the models of group selection. *Quarterly Review of Biology* 53:101–114.

Wade, N. 2009. *The faith instinct: How religion evolved and why it endures.* New York: Penguin.

Warneken, F., Hare, B., Melis, A. P., Hanus, D., and Tomasello, M. 2007. Spontaneous altruism by chimpanzees and young children. *PLoS Biol* 5:e184.

Watson, J. D., and Crick, F. H. C. 1953. A structure for deoxyribose nucleic acid. *Nature* 171:737–738.

Watts, D. P., and Mitani, J. C. 2002. Hunting and meat sharing by chimpanzees at Ngogo, Kibale national park, Uganda. In *Behavioral diversity in chimpanzees and bonobos,* eds. C. Boesch, G. Hohmann, and L. Marchant. Cambridge: Cambridge University Press.

West, S. A., Griffin, A. S., and Gardner, A. 2007. Social semantics: Altruism, cooperation, mutualism, strong reciprocity, and group selection. *Journal of Evolutionary Biology* 20:415–432.

West-Eberhard, M. J. 1979. Sexual selection, social competition, and evolution. *Proceedings of the American Philosophical Society* 123:222–234.

——. 1983. Sexual selection, social competition, and speciation. *Quarterly Review of Biology* 58:155–183.

Westermarck, E. 1906. *The origin and development of the moral ideas.* London: Macmillan.

Whallon, R. 1989. Elements of cultural change in the later Paleolithic. In *The human revolution: Behavioral and biological perspectives on the origins of modern humans,* vol. 1, eds. P. Mellars and C. Stringer. Edinburgh: Edinburgh University Press.

Whiten, A., and Byrne, R. 1988. Tactical deception in primates. *Behavioural and Brain Sciences* 11:233–244.

Whiting, B. B., and Whiting, J. W. M. 1975. *Children of six cultures: A psychocultural analysis.* Cambridge, MA: Harvard University Press.

Wiessner, P. 1982. Risk, reciprocity, and social influences on !Kung San economics. In *Politics and history in band societies,* eds. E. Leacock and R. B. Lee. Cambridge: Cambridge University Press.

——. 1996. Leveling the hunter: Constraints on the status quest in foraging societies. In *Food and the status quest: An interdisciplinary perspective,* eds. P. Wiessner and W. Schiefenhövel. Oxford: Berghahn Books.

——. 2002. Hunting, healing, and hxaro exchange: A long-term perspective on !Kung Ju/'hoansi large-game hunting. *Evolution and Human Behavior* 23:407–436.

——. 2005a. Norm enforcement among the Ju/'hoansi bushmen: A case of strong reciprocity? *Human Nature* 16:115–145.

——. 2005b. Verbal criticism: Ju/'hoansi style punishment. www.peacefulsocieties.org/nar06/060105juho.html.

Williams, G. C. 1966. *Adaptation and natural selection: A critique of some current evolutionary thought.* Princeton, NJ: Princeton University Press.

Wilson, D. S. 1999. A critique of R. D. Alexander's views on group selection. *Biology and Philosophy* 14:431–449.

Wilson, D. S., and Dugatkin, L. A. 1997. Group selection and assortative interactions. *The American Naturalist* 149:336–351.

Wilson, D. S., and Sober, E. 1994. Reintroducing group selection to the human behavioral sciences. *Behavioral and Brain Sciences* 17:585–654.

Wilson, D. S., and Wilson, E. O. 2007. Rethinking the theoretical foundation of sociobiology. *Quarterly Review of Biology* 82:327–348.

Wilson, E. O. 1975. *Sociobiology: The new synthesis.* Cambridge, MA: Harvard University Press.

——. 1978. *On human nature.* Cambridge, MA: Harvard University Press.

Wilson, J. Q. 1993. *The moral sense.* New York: Free Press.

Winterhalder, B. 2001. Intragroup resource transfers: Comparative evidence, models, and implications for human evolution. In *Meat-eating and human evolution,* eds. C. B. Stanford and H. T. Bunn. New York: Oxford University Press.

Winterhalder, B., and Smith, E. A., eds. 1981. *Hunter-gatherer foraging strategies: Ethnographic and archeological analyses.* Chicago: University of Chicago Press.

Wolf, A. P., and Durham, W. H., eds. 2004. *Inbreeding, incest, and the incest taboo: The state of knowledge at the turn of the century.* Stanford, CA: Stanford University Press.

Wolf, J. B., Brodie, E. D., and Moore, A. J. 1999. Interacting phenotypes and the evolutionary process II: Selection resulting from social interactions. *American Naturalist* 153:254–266.

Woodburn, J. C. 1979. Minimal politics: The political organization of the Hadza of North Tanzania. In *Politics in leadership: A comparative perspective,* eds. W. A. Shack and P. S. Cohen. Oxford: Clarendon Press.

——. 1982. Egalitarian societies. *Man* 17:431–451.

Wrangham, R. W. 1987. African apes: The significance of African apes for reconstructing social evolution. In *The evolution of human behavior: Primate models,* ed. W. G. Kinzey. Albany: State University of New York Press.

——. 1999. The evolution of coalitionary killing: The imbalance-of-power hypothesis. *Yearbook of Physical Anthropology* 42:1–30.

——. 2001. The evolution of cooking: A talk with Richard Wrangham. Interview on J. Brockman's website, Edge. www.edge.org/3rd_culture/wrangham/wrangham_index.html.

Wrangham, R. W., and Peterson, D. 1996. *Demonic males: Apes and the origins of human violence.* New York: Houghton Mifflin.

Wright, R. 1994. *The moral animal: Why we are the way we are–The new science of evolutionary psychology.* New York: Vintage.

Zahavi, A. 1995. Altruism as a handicap: The limitations of kin selection and reciprocity. *Journal of Avian Biology* 26:1–3.

INDEX